Developing A Successful Wireless Enterprise Strategy

Developing A Successful Wireless Enterprise Strategy

A Manager's Guide

Scott Sbihli

Wiley Computer Publishing

John Wiley & Sons, Inc.

NEW YORK • CHICHESTER • WEINHEIM • BRISBANE • SINGAPORE • TORONTO

Publisher: Robert Ipsen
Editor: Carol A. Long
Developmental Editor: Adaobi Obi
Managing Editor: John Atkins
Text Design & Composition: MacAllister Publishing Services, LLC

Designations used by companies to distinguish their products are often claimed as trademarks. In all instances where John Wiley & Sons, Inc., is aware of a claim, the product names appear in initial capital or ALL CAPITAL LETTERS. Readers, however, should contact the appropriate companies for more complete information regarding trademarks and registration.

This book is printed on acid-free paper. ∞

This publication is designed to provide accurate and authoritative information in regard to the subject matter covered. It is sold with the understanding that the publisher is not engaged in professional services. If professional advice or other expert assistance is required, the services of a competent professional person should be sought.

Library of Congress Cataloging-in-Publication Data:

ISBN: 0-471-15033-9

Printed in the United States of America.

10 9 8 7 6 5 4 3 2 1

CONTENTS

I've been in the field of wireless and handheld computing ever since 1993, when my software engineering professor walked into the room on the first day of class, held aloft a Newton MessagePad, and spoke these magical words I will never forget, "Who is willing to program this device for my research group in lieu of attending this class and taking the tests? And by the way, I'll pay you for your work." I do believe the speed at which my hand went in the air caused a sound in the room not too dissimilar from a cracking whip.

Although I went to school to become a computer engineer and I'm currently in the computing field, I still believe my life took a 180 on the first day of that software class. I took a day job for five years with an excellent Fortune 50 company out of school, but continued writing and editing for various handheld and wireless magazines in my spare time. One day I woke up and realized that I knew a fair amount about an industry set to grow at a radical rate. So, I started my first company: Empyrean Design Works.

It was shortly thereafter that it occurred to me that my colleagues and I had done a huge amount of work helping companies figure out this space and what to do with wireless computing. We had built expertise and technical know-how. As we continued meeting with potential clients, I found myself spending way, way too much time reprogramming CIOs, directors of IT, and business unit managers. Their heads typically were filled with marketing garbage, lies, and unrealistic beliefs on how wireless computing could or would not help them.

Once I had a director of IT for a Fortune 500 company tell me he wasn't interested in wireless computing because it held no value until third generation (3G) networks become available. He had done his research and was convinced. As you'll learn in this book, wireless is not waiting for high-speed 3G networks; many wireless solutions work just as well without true wireless coverage, and hundreds if not thousands of companies already have real wireless solutions. After one long phone call and one longer meeting with us, he changed his mind.

This book was made to help you change your mind, make up your mind, and maybe even give you your mind (on wireless). We discuss different technologies at a generic and specific level. This book is not intended to push a certain technology, though I do have my favorites, I must admit.

I simply want to educate you about the topics I discuss about wireless time and again with my clients. These are the answers to the questions we get asked the most. Enjoy the book.

In starting a company based on an emerging technology, I've learned a few things. The first is to never assume that what you take to be common knowledge or readily understood actually is. When you eat, sleep, and breathe a technology like wireless you start assuming that everyone inherently understands the benefits of a Palm handheld over a Windows CE device and vice versa. Or you believe that everyone (including their grandmother and the high school kid who bags your groceries) realizes that the current generation of Palm OS-based handheld devices is ill-equipped to handle on-device voice recognition.

Another key to learning is related to the hype, myths, and paranoia that surround any new technology. If you were in the Information Technology (IT) or computer fields during Java's introduction, you'll remember the phenomenal hype that Sun, along with other media, produced with this ubiquitous computing language. One of the predominant claims was that you would be able to write an application once and run it on any computing platform. The virtual machines could execute the program within the limitations of that particular environment. Simultaneously, companies with products competitive to Java blasted it with lists of reasons why it wasn't a language worthy of use. In the end, as with any emerging technology, the truth lies somewhere in between.

So it goes with wireless and handheld computing for the enterprise. On one hand, you will find no shortage of technology and tools companies inflating the abilities of handheld devices to send, receive, and synchronize wireless data. Of course, these companies will throw in ridiculous phrases like, *your data anytime and anywhere*. Great! I would like to see the holographic, three-dimensional cancer fighting protein that my company designed on my Palm at 11 A.M. in my bungalow on a small island in the South Pacific. Maybe that's a bit extreme. Okay, I'll settle for a sales report in a training facility at noon somewhere in a mid-size city. I still can't see my data?

Let's face some facts. True wireless data coverage is poor and it's slow in the United States today. Remember, I said slow . . . not unusable. No less than five two-way wireless data networks within the United States are competing for your company's data. Competition between multiple carriers with multiple standards has, to date, hindered the rollout and coverage of wireless data across the United States. Although Europe and Asia's standards are rapidly becoming more homogenous, they still fight issues of wireless coverage within buildings and throughout a myriad of remote locations, not to mention capacity issues.

Some people will point to third generation (3G) wireless technology as the savior of wireless. Let's be clear on the real and present issues with 3G wireless: cost, competing standards, battery life, and time to roll out among others. In fact, I can make an argument that 384 kbps to 2 Mbps (6 to 35 times faster than a modem) is complete overkill for handheld devices used for business purposes. Even the low end of that range is adequate for transmitting video and audio to a device with a small form factor.

Some may question why someone so obviously negative on wireless technology is writing a book on the subject. I'm actually very positive about the industry and its potential. Even with these challenges, businesses can create real solutions that will result in cost savings, time reductions, increased productivity, responsiveness, and additional customer touch points, as well as new revenue opportunities. Between here (the current state of your business) and there (your business utilizing the benefits of wireless) is a minefield: a minefield of hype, myths, and paranoia.

As mentioned before, hundreds of companies (today) are peddling wireless networks, tools, development environments, business applications, security services, consulting services, development assistance, devices, and wireless middleware. Clients I work with every day express the same concerns:

- What parts of my business make sense to wirelessly enable?
- How will wireless and handheld technology impact my business strategy as well as product and service lines?
- How do I know which companies' technologies to choose?

- Can my enterprise rely on the fact that these companies will be around in one year? Three years?

- With so many wireless technologies running around, are there any true standards I can rely on?

- What is the cost to deliver wireless solutions?

- How do I prevent the wireless infrastructure I choose from becoming obsolete in 18 months?

These questions are one of the largest reasons I wrote this book. I abhor misinformation. Little information is published today about true enterprise wireless strategy and solutions. By watching television or reading advertisements from product companies, you would think differently. The cellular phone companies would make you believe that all wireless data will appear on your phone's display and that you would enjoy entering information using a 12-button keypad. Are they kidding?

Other companies boil wireless solutions down to pumping out Web pages to handheld devices. Did anyone mention to them the costs and lack of availability of wireless coverage in most parts of the United States? What about the fact that Web solution today has limited user interface designs? User interface design may be one of the most important reasons your enterprise or customer solution will succeed or fail. With a small screen, you must be creative and think outside the PC paradigm.

Enterprise wireless strategy and solutions are not simply about 2G and 3G cellular networks, cell phones, Web browsers, and email. Strategies we conceive, technologies we assess, and solutions we build are concerned with subjects like business process redesign, store-and-forward synchronization, security of data on devices and through the air, wireless middleware options, synchronization business logic, cross-platform applications, and electronic software distribution.

My hope is that this book educates you on the real issues of developing enterprise wireless strategy and solutions. I want it to bring to light the issues we work through with clients day-in and day-out independent of their industry. I also would like you to be armed with the information that will help you filter out the junky or parity technologies that some companies hype. You'll be able to wrap your mind around the levers you can pull to build your solution and the questions to grill a vendor on. This book is for your education. This book is for your success.

Overview of the Book and Technology

This book will also guide you safely through this deadly minefield, and do it quickly whether your field is automotive, healthcare, utilities, financial services, field force, sales force, or any other industry that can reap the benefits of wireless and handheld technology in the enterprise. Your competition is not waiting for you, and the cycle time from when a technology is introduced to when it is embraced by business continues to shrink.

Your enterprise needs to understand the pros, cons, and intangibles of wireless to give you a competitive edge. This book will guide you through developing a strategy for your business. You must understand what to enable, what *not* to enable, and how to enable it. A successful wireless strategy and implementation requires the knowledge to answer the questions posed previously as well as knowledge in synchronization, project planning, application development, architecture design, strategy, and devices, among others. We'll cover all of these topics throughout the book to ensure your success with a technology set to revolutionize business.

I'm a big believer in examples and illustrations. It helps to take nebulous topics and ideas and make them concrete. I once met with a client in the railroad industry who knew he needed a sales force automation tool for his sales team. Conceptually, he understood the benefits. When my colleague showed a generic version of a tool we developed for another client, his eyes lit up. The connection between theory and practice was made. This book points to examples, names specific companies, and sometimes lists prices of certain items. Although some of the information may become out of date over time, I strive to write something that is practical to the enterprise manager.

However, this book is not going to explain programming or show it, dig into wireless protocols and argue their pros and cons, or speak on how to install synchronization servers. This book is about business strategy, technical strategy, and implementation guidelines. The book will be littered with examples and will do its best to explain different technologies in layman's terms. As when I taught my students at the University of Michigan, my goal is to have the reader read the book, put it down, and utter, "I get it!" So, get to it . . .

How This Book Is Organized

Knowledge is an island (see Figure I.1). You may learn information about handheld devices. You may learn other information about wireless networks. Finally, you may become an expert in database technology. Each of these blocks of information is an island. The key is how these islands are connected. How does a handheld relate to a wireless network and which ones can it talk to? How does a wireless network facilitate the transfer of data between a handheld device and a corporate database? Does the handheld talk directly to my database and what happens to that conversation should I lose wireless connectivity? These bridges represent the connections between the knowledge we collect throughout our lives. This book follows that approach.

We will explore all of the key components of handheld and wireless computing *and* build the bridges that connect these pieces. If that were not enough, we'll hop in a plane and give you the 50,000-foot view of these islands and bridges. From this viewpoint, you can evaluate the

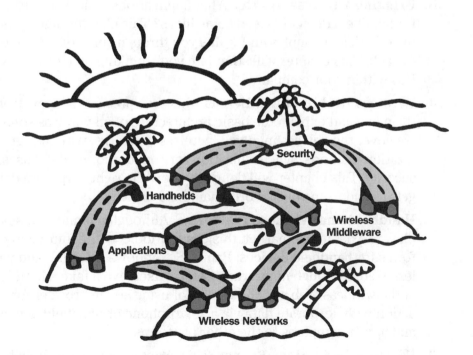

Figure I.1 Islands of information and interconnections.

pieces important to your enterprise and as well as formulate what to wirelessly enable and how to create the ROI your business needs.

To accomplish our goal, we've laid the chapters out in the following order:

1. **Wireless As a Business Advantage.** This chapter lays the groundwork for wireless. It explains the forces that have made wireless and handheld computing so incredibly popular. It answers questions on the business drivers for such solutions by describing those types of applications and subsets of information that can be enabled through wireless technology. It also helps your enterprise understand why it is a competitive advantage to embrace this technology today.

2. **Developing a Wireless Enterprise Strategy.** No technology should be put in place for the sake of the technology. What is the business justification or ROI for wireless and handheld solutions? We'll cover the impact on products and services, your company's business processes, and technology architecture. This chapter will also help you create a value map of wireless solutions for your company and help you organize a priority list.

3. **Bringing Wireless to Life.** What applications make sense to create for handheld devices? Need some ideas? This chapter highlights four hypothetical examples and walks you through some of the benefits of each. This chapter will stimulate your thinking on ideas for wireless within your company.

4. **Wireless Architecture.** Here is where we present the view from the plane and explain the basic architecture of all wireless solutions. We cover the devices involved, wireless networks, wireline synchronization methods, synchronization servers, corporate databases, and security. This chapter will demystify how the right corporate data gets to and from your favorite handheld device.

5. **Handheld and Wireless Platforms.** Although a number of smaller players exist in the market, this chapter looks at the four main platforms for handheld devices: Palm OS, Windows CE, RIM, and wireless Web. Although not a true platform itself, we'll take a hard look at the wireless Web and the virtues of using a Web browser for interaction with corporate data via a smart phone or handheld device and how it compares to the other platforms.

6. **Wireless Technology Primer.** This chapter tackles the alphabet soup of wireless technology. Delivered in layman's terms, you'll gain

understanding of such terms as CDMA, TDMA, Bluetooth, and 2.5G. We'll explain the differences between wide area and local area networks. By the end of the chapter, you'll know where wireless has been and where it's likely heading.

7. **Wireless Middleware, Leprechauns, and a Magical Rainbow.** For the most part, the magic of data synchronization doesn't happen on the device itself. Rather, building business logic to support synchronization across multiple device types, data sources, connection types, and coverage levels requires extensive forethought, design, and modification of existing business systems. Learn how wireless middleware works, what the challenges of synchronization are, and how to deal with those challenges in all types of environments. More importantly, learn what leprechauns and a rainbow have to do with synchronization.

8. **Handheld Application Design.** A battle rages in the hearts of men and women between writing applications for handheld devices and just delivering the content via a Web browser. Maybe that's a bit dramatic. However, this chapter discusses many design issues, including native applications, Web-based applications, interface design, Java, and Rapid Application Development (RAD) tools. Determining how to build your solution will depend on what horizontal or vertical market you're in, who your customer is, and when you build your solution. We'll also explore solutions for delivering applications that you write once and can distribute to any type of handheld or wireless device.

9. **The Cost To Go Wireless.** This chapter covers all of the costs associated with wireless and handheld computing. Project costs, hardware, software, wireless data, consulting services, and synchronization middleware all play into the picture. Should you host this infrastructure yourself or outsource? Get a handle on costs before you start your project.

10. **Wireless Case Study: Ordering and Inventory.** Follow along on a case study of a business-to-business dot-com that wanted to extend ordering and inventory services to their customers. Go through the strategy, design decisions, architecture, technologies, rollout, and results of this large-scale wireless implementation to understand how business processes were streamlined and costs reduced.

11. **Wireless Case Study: Healthcare Charge Capture.** The second case study looks at an area of healthcare that causes doctors and hospitals to lose huge amounts of money and a handheld application that helps recover those dollars. MD Coder is an application for charge capture in healthcare and represents a good example of enabling doctors at the time and location they need to collect information the most.

12. **Wireless Technology Trends and Conclusions.** This chapter presents a short- and long-term forecast for this industry. We'll look at devices and convergence; we'll explore the future of wireless networks including 3G technology for a cure to our wireless blues, and look at the future of the wireless architecture presented in Chapter 4. Will thick clients or thin clients prevail? What role does the eXtensible Markup Language (XML) play in the future? What about biometrics for security?

I've written this book to demystify the strategy of implementation of wireless solutions. This book will give you the 50,000-foot view, the islands of information, and the connections between those islands. Whether you decide to enable these solutions yourself or use outside resources and expertise is up to you. Either way, you'll be able to make your decisions in an informed manner and will never be at the mercy of product marketers.

Who Should Read This Book?

This book is written for a variety of audiences independent of its title and I believe everyone from the CIO to the IT project manager can reap huge benefits from this book. By and large, this book's angle is of the business perspective with technical information where it is needed to explain the concepts for developing a wireless strategy.

For the CIO or IT director, this book will lay out all of the aspects of wireless computing from a strategy, process, and deployment perspective. You'll quickly get your arms around those business processes and functions that make sense to enable. It will help you to prioritize your wireless offerings and solidify expected costs and staffing. If you choose to take a deeper dive into technology to understand the networks, synchronization middleware, and platforms, you have that option, as well.

Project managers will gain a keen understanding of the overall wireless architecture and technology tools. It will help him or her filter out the technology garbage in the industry and understand the intricate challenges of designing handheld applications and synchronization code. A project manager who reads this book will greatly increase his or her chances of a successful wireless implementation.

This book makes few assumptions about your technology background other than that you have some IT experience and understand the fundamentals of PC-computing and networking. This book is *not* going to instruct you on how to develop Java, C, or Web applications for handheld devices. Plenty of other resources are out there for achieving those goals. *Developing a Successful Wireless Enterprise Strategy: a Manager's Guide* is here to lead you through the strategy, technology, and management issues of creating successful wireless solutions for your business. Enjoy.

CHAPTER 1

Wireless as a Business Advantage

Executives, marketers, product managers, and developers jump on technology trends at a rate that becomes more alarming each year. Whereas it took three, four, or even eight years for some technology trends to take hold and mature to a point where an enterprise could embrace them, companies are expected to make the same decisions in 12 to 18 months today.

A decade ago, when mainframe computing was a hot topic and Enterprise Resource Planning (ERP) got its roots in the enterprise, the rate of growth of particular computing trends and their adoption by enterprise was slower than the trends of today (see Figure 1.1). Barely a decade later, several more evolutions or revolutions, depending on your perspective, have happened. More importantly, each curve in the graph has a steeper slope than ones previous to it indicating a relative growth rate of its popularity and adoption. The frenzy to retain competitive advantage forces many enterprises to embrace new technologies at a faster clip. Although Information Technology departments have to discern fad from trend, there is no doubt that the cumulative evolution of computing infrastructure and trends leads to new trends that evolve and are adopted at a quicker rate.

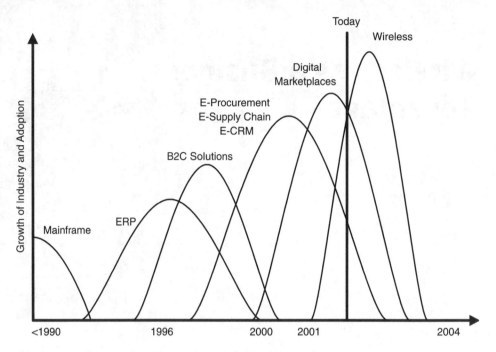

Figure 1.1   The growth and adoption of computing trends.

Wireless computing is currently the cream filling in the technology donut. Believe it or not, wireless has many meanings depending on whom you ask. I can tell you that wireless computing today involves more often than not wires, cables, and cradles. As we'll learn throughout the book, wireless data connectivity is in its infancy, and coverage is not guaranteed. So, I'll use different terms: wireless computing and handheld computing. I'll refer to handheld devices, handhelds, and Personal Digital Assistants (PDAs). Whatever the exact term, we're talking about the class of computing that Palm, Microsoft, Compaq, Psion, Nokia, Motorola, Ericsson, and others have popularized in the past two or three years. I'm also referring to software and hardware solutions that have the capability to be wireless, but for various reasons enterprises choose not to roll them out as wireless to begin with.

That said, we are on the cusp of a technology and business trend that may be as impact filled as the Internet itself. Its preeminence stems from the fact that it impacts businesses and consumers alike on a global basis. Wireless computing, mobile computing, handheld computing, or whatever else you choose to call it has the capability to redefine computing because it brings with it a certain ubiquity of information. How will businesses capitalize on it or at a minimum use it to not lose precious ground to the competition? How will enterprises choose to market products and services to consumers who desire information and the ability to transact business from any location? Finally, which completely new revenue streams, cost savings, and productivity improvements will be created because information can be delivered where and when it has the most impact?

Business Advantages of Wireless

So, let's start with a basic business question: What is the business advantage of creating wireless products, services, and solutions whether it's for a client, customer, consumer, or internal purposes? Really, it has to come down to one of four things:

- Cost reductions
- Productivity increases
- New revenue opportunities
- Enhanced information and decision making abilities

Arguably, you could say it comes down to one factor: dollars. How will my company save money (reduce total dollars spent), become more productive (increase transactions per dollar spent), find new revenue opportunities (increase total dollars), or provide more knowledge (customer or consumer is more likely to spend dollars with relevant information) through the use of wireless technology? If you can't answer this question, there is little reason to proceed. However, I encourage you to read on for two reasons. One, you likely just paid for this book and want to get your money's worth, and, two, we'll cover many examples and case studies to jumpstart your brain and fill it with wireless solutions ideas for your enterprise.

Time and Location Services

The good news is that there are a multitude of solutions you can create using wireless and handheld technology. In fact, new revenue streams will present themselves, and customer touchpoints will be created because of this technology. Wireless technology will evolve into ubiquitous computing opportunities. Those types of applications that are most applicable and potentially profitable to wireless computing are those that contain information that is time and/or location sensitive. Because you are not restricted to data access while sitting in front of a computer at a desk, you can begin to think of your data in all new ways. You can now look at your employees' or customers' business process and determine what information they will need at different times and different locations.

Location Sensitivity

Location sensitive information and business processes deal with having access to information and being able to take action because of a need at a particular location. The ability to order a book from Amazon on a handheld is arguably location *insensitive*. It is unlikely that the book you are ordering is a real need and not a want with location convenience.

There are two subtypes to location sensitivity. First is access to information and systems at a particular location because that business process takes place away from the office, the computer, and the standard corporate LAN. A field force worker who delivers water at different stops along a route has need for information because his business process is in the field. The second type is access to information *because* you are away from the place you normally do business. A manager on a business trip may need access to email or need the ability to submit time and expense reports while out of the office. Both of these situations qualify as location sensitive.

Field forces, such as delivery services, plumbers, electricians, sales teams, and home healthcare workers, are all examples of people who need to either collect and submit or have access to corporate information while in the field. Their need to obtain access to or submit that information at a particular *time* may not be as important. Delivery service workers need to know what to drop off, what to pick up (potentially), what their next stop is, and what information is necessary for invoicing. This all happens on location at the customer's business or residence. There is no need to collect or view information at a particular time, necessarily.

Home healthcare workers, for example, also collect large quantities of information for reporting and billing purposes. When they visit a patient at their homes, they are responsible for recording vital signs, cognitive abilities, and supplies used while onsite. They may also need to reference information while in a patient's home such as insurance policy numbers, physician phone numbers, or a medication list. In both examples, the *time* this information is documented isn't as crucial as the need to record it at a particular *location*.

Location Sensitivity Example: Sales Account Manager

Let's look at a more in-depth example of a location sensitive application and how wireless and handheld technology can create business opportunities. A sales account manager steps into her customer's office for a customary visit. She asks the customer if the parts he has been receiving have been on time and up to the quality levels he expects. Her customer responds with an affirmative and also wants to know if an order can be placed for 20,000 additional parts due to an inordinate consumer demand for the product.

The sales account manager using a wireless handheld device retrieves customer account information while simultaneously submitting the customer's inquiry for 20,000 parts for the product (see Figure 1.2). The request finds its way back to the company network and eventually off to manufacturing to check inventory levels, line capacity, and scheduling ability. This request ties together information from a number of different systems. Due to the time criticality of the request, the sales account manager must charge a premium price due to overtime hours necessary for manufacturing. However, she presents a fixed and known cost to the customer while smiling and saying that her company can fulfill the request without issues.

Her customer accepts the proposal and the order is placed. The submitted order triggers a number of processes within the supplier's back office including the alert of manufacturing to the request and accounts receivable to book the revenue and begin the billing process.

Unlike any time in the past, this sales account manager is able to request information for her customer as well as book the order while sitting in his office. The location sensitivity of this request and sales

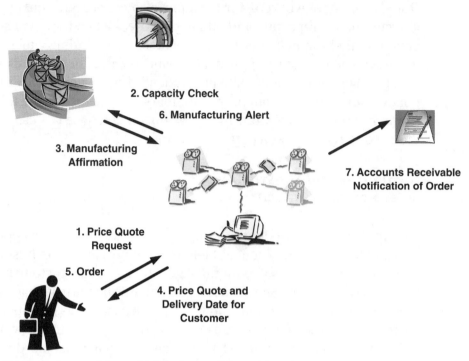

Figure 1.2   A workflow with location sensitive information.

manager's ability to fulfill it is a win for her company. Her client is ecstatic that no follow-up phone calls are needed or potential delays added that would prevent him from supplying an additional 20,000 items to meet consumer demand. A traditional sales request would have resulted in the sales manager needing to the leave the facility, make phone calls to different departments to verify that her company can fulfill the order, and enter the booked order into an order entry system back at the office. Handheld and wireless technology in our example mitigates the need for many of these steps and leaves the customer satisfied to know the fulfillment process is underway and demand for his product will be met.

Location Sensitivity Example Advantages

A plethora of benefits result from the use of wireless and handheld technology at the *location* they are needed:

Enhanced customer information. The size and portability of hand-held devices make them ideal and unobtrusive for referencing customer information and placing orders. Before today, it would have been difficult if not impossible to arm the sales account manager with the ability to place orders on the spot, research prior sales quantities for the customer, present detailed or technical information on all products or services, and dynamically update customer contact information.

Time to delivery. Orders placed onsite speed the date to delivery. If your company plays in the services industry, placing an order may mean searching a resource database for skilled workers and scheduling them to be at a client site the next day.

Shortened cash cycle. When the order is placed, accounts receivable can be notified to begin the billing process, shortening the time from when the customer gives approval for an order to when monies are collected.

Information technology benefits. From an IT perspective, the cost to support a handheld and wireless infrastructure is less than a laptop infrastructure. In a later chapter, we'll cover maintenance and remote updates of handheld devices.

Real intangibles. The handheld device enhances the experience at the customer site as noted previously and replaces the laptop that most sales representatives would never use at a customer site. Laptops are bulky, have a long boot time, but most importantly construct a physical barrier between the rep and the client, which diverts significant attention away from the conversation.

Obviously, this sales scenario will not apply to all enterprises, but it does for a great many. If your order placement process differs from this, take heart. You'll learn to map your business process to the handheld device and gain similar, if different, advantages.

Also, it would behoove us to stop for a moment and recognize the complex, but realistic business processes that were initiated in the course of the sales account manager performing her job. The request for a quote resulted in multiple back office systems talking with each other to determine if sufficient inventory exists, the calculation of an overtime premium on the order, and the notification to manufacturing and accounts receivable of the order itself. No longer are you relegated to simple pages, emails, and phone calls to undertake processes that can be automated for timelier customer response. Such integration does require more work (strategy, business process redesign, development, integration, and rollout) than handing out wireless devices and a Web browser to the entire sales force.

Time Sensitivity

The other important factor that many times determines the need for wireless or handheld computing is time sensitivity. Time-sensitive information is data that is needed at a precise moment of the day (such as 5:30 P.M.) or at the exact moment it is available (such as a change in inventory levels below a particular threshold). In the latter example, we typically call these types of events *alerts*. All data has some level of time sensitivity, whether it's low or high. Table 1.1 highlights a few vertical markets and processes that possess time-sensitive data.

The concept of being able to notify people with time-sensitive information wherever they are is not new by any stretch. One-way paging networks have been in place and in use by doctors, IT support people, lawyers, field service workers, and other groups for many years. Recently, many of these telecommunications companies have imple-

Table 1.1   Examples of Time-Sensitive Data

AREA	TIME-SENSITIVE DATA
Healthcare	Lab results—Doctors and other healthcare workers need to be alerted with patient lab results the moment they are available.
Transportation and logistics	Routing—A driver may be required to add a stop to a delivery route if a customer calls in to the dispatch office while the driver is on the route. In this case, the driver is notified of that stop, what products are required, and can acknowledge the delivery once there.
Financial services	Stock trading—Traders need to know when certain prices go above or below set thresholds and to take action on those changes at a moment's notice.
Manufacturing	Line operation—Workers at a plant need to be notified when an assembly line goes down and also to find additional details out as to why the event occurred.

mented two-way paging networks. Not only can you receive a page with a return phone number or text message, but on a two-way network, users can take action on the requests by typing out short messages or choosing a response from a predefined list.

Therein lies the key to time-sensitive information: having the ability to be notified of a change or with a piece of information *and* the ability to *act* upon this information. Handheld devices and smart phones excel at this capability. With included memory, processing power, a large-ish screen (relative to pagers), and an operating system to manage resources, users have the opportunity to not only respond, but

- Trigger a business workflow to deal with the situation or information.
- Use the processing power of the device to drill-down into details, receive additional information through requests and alerts, and use computing power to process the data or change the view into it.
- Engage others in the situation based on the content of the information.

Healthcare is a field in which data is critical at a particular location and time. In this example, we'll look at the time sensitivity of the situation.

Time Sensitivity Example: Doctors

An orthopedic surgeon sleeps at home while a nursing home patient collapses on the way to the restroom breaking her hip. The doctor at the nursing home has an ambulance sent to take the patient to the hospital. The hospital notifies the sleeping orthopedic surgeon by paging a wireless device (see Figure 1.3). The waking doctor reads the patient chart noting vital signs, ETA to the hospital of the ambulance, and the fact that surgery will likely be needed. The device also lists two available operating rooms. The doctor responds that he received the information before dressing and driving to the hospital.

After the patient is stabilized by ER doctors, she enters surgery and has her hip successfully repaired. The patient is then sent to a room for recovery. The orthopedic doctor visits the patient on his rounds a day later after viewing the chart of the patient from his handheld device. While visiting the patient, the surgeon notes redness, hotness, and swelling around the surgery area suggesting an infection of some type.

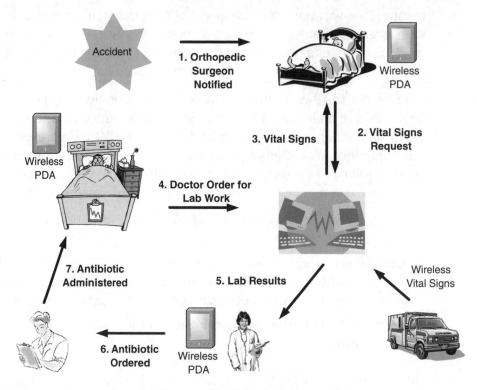

Figure 1.3 A workflow with time-sensitive information.

The doctor orders a blood test from the wireless device. This notifies the nurse to draw the blood, alerts the lab in advance that their services will be needed, and updates the patient's medical chart.

Six hours later the results are known and keyed into the hospital's patient tracking system. By then the orthopedic surgeon has gone on to the next hospital to see other patients. With a wireless device at his side, there is no problem notifying the doctor of the results. After interpreting them, he orders that an antibiotic be administered. This action causes an electronic prescription to be sent to the hospital's pharmacy and alerts the nurse to administer the antibiotics per the doctor's instructions.

Time Sensitivity Example Advantages

In this example, we note a number of efficiencies that are realized from the use of handheld and wireless technologies for patient care:

Informed decisions. The orthopedic doctor, when paged, has copious amounts of information available to him including current vital signs, status, ETA to the hospital, medications, allergies, and other pieces of patient history. Typically a doctor talks by phone with the hospital to piece together a picture of the patient's condition.

Information accuracy. If the doctor has available to him an electronic patient history from a central location, fewer mistakes will occur in caring for the patient. These miscues may arise from lack of information (such as allergic reactions) or mistakes due to improperly relayed information that can occur from multiple people in a communication chain.

Reduced hospital costs. Without a doubt, hospitals have numerous information systems (both electronic and paper) and communication chains. Patients, doctors, nurses, and other healthcare workers spend inordinate amounts of time waiting for data or information that is already known by someone else. Pages, phone calls, and the physical transfer of charts can be lessened through centralized data stores and mobile computing. Cutting time will ultimately save the hospital money and enable them to see more patients per year.

Improved patient care. No part of the previous scenario has life or death consequences. However, a patient will receive improved care because of more timely information. The surgeon had charts and patient history at the time he was paged, reducing the chances for mistakes. Immediate lab results enabled a patient to begin antibiotic treatments in a more timely manner lessening the total time she was in pain.

Both Time and Location

All business processes have some combination of time and location sensitivity. In the sales account manager example (previous location example) there are portions of the business process where you can point to time sensitivity of information as well as location. The fact that the sales manager could place the order on the customer's premises allowed her to take advantage of the sale at that moment. Had she told her customer, it would take 24 hours to get a quote, the customer could easily have sought out his secondary supplier of parts.

Similarly, the orthopedic surgeon (previous time example) works in numerous locations; there are location sensitive portions to his work.

He is paged at home. He visits hospital ERs, patient rooms, and may make house calls. His work takes him to nursing homes and potentially three or four hospitals. The nature of the job makes the information he needs necessary on a location as well as a time basis.

So, there is overlap. Many handheld and wireless applications have both aspects of time and location sensitivity. The key to remember is that they must contain at minimum one of these aspects to justify their existence in most cases. Getting wireless email if you work in a cubicle all day is worthless. The job has no need for location or time-sensitive information. As we continue on through the book, you'll discover that not all applications with one or both of these criteria make sense to create on a handheld or wireless device either. Time and/or location are necessary, but not necessarily sufficient for finding the low-hanging fruit you may be seeking.

The Drive to Wireless

Now that you have a basic understanding of the business drivers for wireless technologies, let's spend a bit of time understanding how wireless computing came to be and more importantly why today is the right time to build your company's wireless strategy. Some may think this backwards: trying to show the business case before explaining how we got here in the first place. I believe in practical examples. The previous business examples likely have you pondering your own business and maybe a light bulb has gone off in your head regarding particular areas of your business. However, there are good and logical reasons why wireless computing is such a hot and viable area for business today.

The Internet

The Internet is the first of four forces that have set the stage for wireless computing. Arguably, you can say it is the most influential. The Internet has redefined how we live, learn, and communicate. It reaches all over the world and there will be a day in the not too distant future that the vast majority of the world's population will have easy access to this network. Its evolution has shifted our thinking on information availability and communication.

A few short years ago, a high school student would need to trek to a library to have access to books and encyclopedia articles about France, for example. Contrast this with today's student who can sit in front of a computer screen and not only read several different encyclopedia articles on the subject, but read English and French versions of the articles, visit thousands of Web sites hosted in France, watch streaming videos about France, and build a friendship in France via an instant messaging system. Not only can the student report on the country, but he can also bring in the personal perspectives of the people living there.

Initially, the Internet was text based and available only to the government and university researchers. However, that has evolved over the years. Today's user surfs through graphically rich Web sites with synchronized audio; he experiences content and opinions of people all over the world and does this through a high-speed connection. The next step in the evolution is to retrieve this information while in a disconnected or wireless mode.

One example of the Internet reaching us wherever we go is the Kyocera smartphone. In mid-2001, Kyocera released a device that represents a second-generation convergence wireless communicator (see Figure 1.4). It combines the wireless communication of a digital cell phone with the data capabilities of one of the most popular handhelds, a Palm OS-based device. This device has the computing abilities to handle everything from the generic address book, to-do list, and calendar, to specific business applications such as sales force automation or route-based delivery. The Web browser works wirelessly and comes with high-end features such as Caller-ID, voice dial, and text messaging.

Wireless Networks

Wireless networks have been around for a number of years and, believe it or not, have made monumental improvements over the past few years. The United States has been making rapid additions to its networks, converting them from analog (first generation) to digital (second generation), albeit with three digital standards: CDMA, TDMA, and GSM. Europe and Japan, partially due to their smaller landmasses, are getting their first taste of 2.5G, (two and a half generation) higher-speed networks.

Figure 1.4 The Kyocera QCP-6035.

The wireless networks that the United States has are all more or less capable of sending data in both directions although at low speeds, 819.2 kbps. These speeds represent one-seventh to one-third of the speed of a conventional PC's landline modem, but are absolutely usable for business applications on wireless devices.

As we will explore in more detail in an upcoming chapter, there are a number of viable networks available to your business. Some cities are covered by five or more while others may have only one. The networks mentioned previously are all based on cellular technology. Additionally, you have the option of choosing from a number of one- and two-way paging networks as well as data-only networks.

Consumer and business demand for data services will continue to drive the rapid build out of these networks. Today it is estimated that almost 40 million people in the world use wireless Internet access, or more generically, send and receive data through a wireless connection. That is expected to grow to almost 730 million people by 2005 according to *mBusiness* magazine's April 2001 issue (see Figure 1.5).

Computing Power

Moore's Law is often quoted in books, magazines, and Web sites. Gordon Moore, former chairman of the Intel Corporation, stated in the 1960s that a computer's processing power will double every 18 months. That insight has held true since he made the statement some 30 plus years ago. Each year chip manufacturers find ways of shrinking the fabrication process of microprocessors. This results in the creation of chips with more computing power per square inch of silicon, higher clock speeds (commonly called Megahertz), and lower power consumption needs.

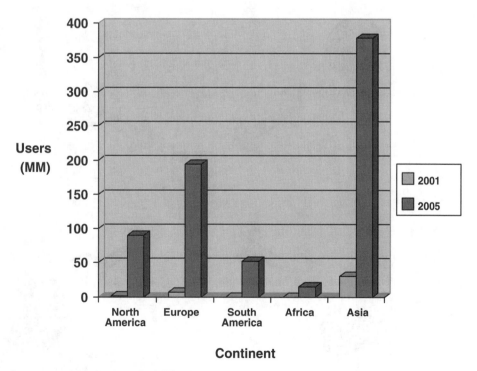

Figure 1.5 Wireless Internet (data) users.

Those factors also enable a single chip to perform more functions. For example, a first generation PDA probably had separate chips for processing data, handling the screen display, and controlling any input and output ports on the device. Subsequent generations of the same device found all of those functions controlled by one smaller chip. Today's handheld devices are beginning to handle images, video, audio, processing, and wireless communication from one or two chips. It is these types of improvements that enable each of us to carry devices around that are capable of handling complex business problems, and also to tuck them away in a shirt pocket or purse.

Social and Economic Forces

Lastly, social and economic forces also play a significant role in wireless computing (see Figure 1.6):

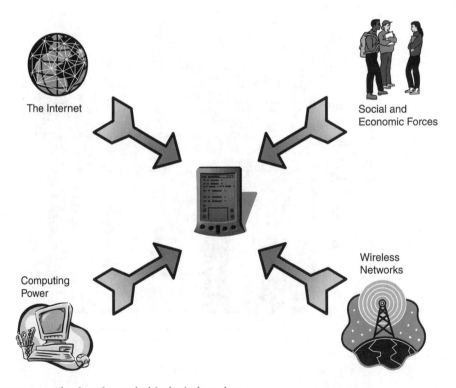

The Internet

Social and Economic Forces

Computing Power

Wireless Networks

Figure 1.6 The four forces behind wireless data.

- **Flat out, people are becoming information junkies.** The sheer volume of information on the Internet, along with the low barrier to entry, means that people expect to have access whenever they want it. Going on vacation? Search the Internet for recommendations or even look out over the Bahamas through a Webcam. Want a recommendation on a car before visiting the dealership? Search the Internet for reviews. Want the most up-to-date weather forecast? Search the Internet for a real time radar animation.

- **The reduced cost of computing power.** Free cell phones and free computers (provided you pay for Internet access) continue to lower the barrier on wireless communication and information access. There will be a day in the not too distant future where every person with a cell phone has Internet data access capability.

- **The mobility of people.** It seems as if the world is shrinking to the size of a globe. At one time Europe was a semi-exotic place due to the cost and time it took to get there. Multiple languages in a small geography made it a challenge to get around. Today, $200 airfares make Western Europe a spot for a long weekend. Handheld translators help break down language barriers. As the opportunities for travel continue, consumers and business people will demand information wherever they are.

- **There is a dramatic shift in cellular phone access from landline.** Japan recently saw the tide turn in 2000 when cellular phone subscribers exceeded landline subscribers. In other countries, particularly third-world countries only now installing a communication infrastructure, communication companies reap the benefits of installing a cellular only network avoiding the costs of laying copper or fiber optic landlines.

The Bridge from Here to There

In this beginning chapter we looked at some of the most fundamental issues related to wireless data and communication. We answered the question of why people are clamoring for wireless data (computing power, the Internet, wireless networks, and social and economic forces) and why telecommunication and technology companies are scrambling to fill that gap.

More importantly, we took the next step and answered the question of what role wireless or handheld data plays in an enterprise. Enterprises should search for applications where information is necessary in time or location sensitive situations. We presented two hypothetical, but common situations and demonstrated qualitatively why there is a business advantage to using wireless and handheld computing. This chapter likely answered as many questions as it generated.

While there are plenty of technology concepts related to wireless to get through, we're going to spend the next couple of chapters continuing to look at strategy and examples. Without making the proper case for what you are doing with wireless, you will get little buy-in and potentially much resentment. With the proper strategy and its tie-in to overall corporate goals, the benefits will be many.

CHAPTER 2

Developing a Wireless Enterprise Strategy

Ah, strategy. Assuming your company has an overall business strategy, and perhaps even a technology strategy, it would seem to make sense that any company even considering wireless should develop a strategy for wireless technology. What do we mean by strategy? How do I go about creating one? Who in my organization should be involved?

This chapter sets us off to answer these questions. We will start by defining a wireless enterprise strategy and describe how this strategy should link to your overall business and technology strategy. We will then turn our attention to the importance of developing a wireless strategy before getting too far down the mobile implementation path. This will quickly lead us into a step-by-step review of the critical components of a wireless strategy and the process for developing one. Finally, we will briefly consider the idea of prototyping and its relationship to strategy and discuss the team you will need to make this all happen. Hopefully, by the time we are done you will not only understand what we mean by a wireless enterprise strategy, but you will be well on your way to creating one for your company.

Wireless Strategy Defined

Strategy can mean numerous things to different people. At the highest levels of the enterprise and in academia, strategy generally refers to the set of techniques a company will embrace to operate the business, grow the company, and beat out the competition. Under this definition, strategy can include

1. A definition of the products and services your company will offer.
2. The pricing for your products and services.
3. How you plan to bring the products and services to market (your marketing channels) and embrace your customers.
4. Anything else that helps to explain how your company will win in the marketplace and wipe out the competition by differentiating itself.

There are literally hundreds of books on the market today that present some variation on the previous definition. Let's assume for our purposes that the previous is a decent cut at a definition for business strategy. Are we ready to describe what we mean by wireless strategy? Not quite yet. Before we can apply this definition, we must first understand the concept of an *enterprise technology strategy*.

Technology marches forward, becoming a more significant part of most companies' operations each day. Of course, it also continues to become a larger part of the cost budget. This has made technology spending visible at the highest levels of the company and with it the need to plan carefully for this spending. We have all seen companies that have not done an effective job. We've witnessed several recent announcements of companies spending hundreds of millions of dollars on Enterprise Resource Planning applications (ERP) or supply chain systems only to stop their deployment midstream. A well-defined enterprise technology strategy can help ensure that technology resources are deployed logically and in support of the company's business strategy. The weapon of choice: the information technology strategic plan. This document captures a company's business strategy and describes the plan for using technology to effectively support this strategy.

So what do we mean by an enterprise wireless strategy? Simply put, it is the set of techniques your company adopts for deploying wireless technologies in support of your overall business strategy. Your wireless strategy should be discrete from, yet linked to, your business and technology strategies (see Figure 2.1). Think of these three areas as a pyramid, each linked to the other two and influencing the others to some degree. Business strategy is at the top of the pyramid and is clearly the dominant component. Yet technology strategy and wireless strategy can influence business strategy. Consider for example the company that develops a wireless strategy and in the process determines an efficient way to reach new customers. The company decides that this new customer group is extremely valuable and shifts the focus of the company to these new customers, even changing its product and service offerings. Clearly, in this instance the company's wireless strategy had an impact on the overall business strategy.

For our purposes, we will define a wireless enterprise strategy fairly broadly. By this we mean that the strategy will cover the potential for

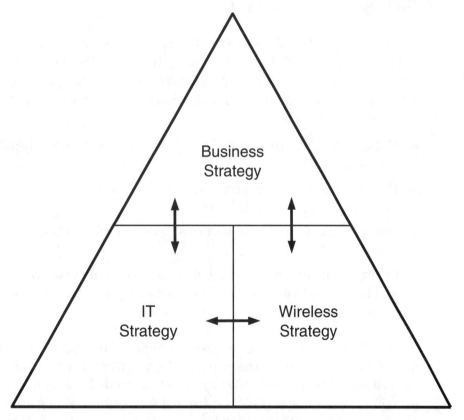

Figure 2.1   Relationships between business, technology, and wireless enterprise strategy.

wireless solutions for the company and the plan for implementing these solutions. In other words, our strategy will include the creation of an operational plan for making the strategy happen. As more than one Fortune 500 CEO has stated, strategy disconnected from operations is meaningless, for it is through business operations that your strategy comes to life.

Why Create a Wireless Enterprise Strategy?

So now we must ask ourselves, "Why is this necessary?" It may seem logical to simply begin implementing wireless solutions quickly so that your company can gain a business advantage. There are at least three compelling reasons for building a wireless strategy for your enterprise before you do anything else.

1. No Strategy Is a Strategy

As the rock band Rush once said, "if you choose not to decide, you still have made a choice" (Rush, "Freewill," from the album titled *Permanent Waves*, 1980). This definitely applies to developing a wireless strategy. Look around most organizations today and you will see some or all of the following:

- Mobile phones in the hands of most executives and nearly all field personnel
- Palm OS and Pocket PC handheld devices in use for personal information management (PIM) activities
- Selective use of real-time wireless devices such as Palm VII handhelds
- Other mobile and wireless technologies, including two-way pagers and Research In Motion (RIM) devices for wireless connectivity to corporate email

If your company has mobile workers, chances are a significant percentage of these are using wireless technology in their day-to-day business operations. This use will only skyrocket as the technologies mature. Do you think your company has sensitive information walking around on employees' handheld devices? You can almost guarantee it does. I can tell you from personal experience that people become extremely attached to a specific brand or type of device. Your ability to deploy mobile applications in the future could be significantly more difficult if you have several device types and operating systems in use versus only one or two.

By choosing to not develop a wireless strategy for your company, you are essentially adopting a strategy of allowing individuals, groups, departments, and divisions to do what they want in the mobile computing space. This strategy will likely include some interesting components: standards don't matter, security is up to you to manage, and everyone picks their favorite devices and applications. Not a pretty picture to be sure.

2. Leverage Your Investments

Let's say you decide to build and deploy a mobile application for your field workers. You choose what you think is the perfect device and infrastructure and off you go. Six months later, you discover the need for another application for this same group of workers. Unfortunately, this other application requires a different device and infrastructure, which could have been used for the first application as well. What to do? Throw out the old investment and start over? Put off implementation of the new application until you depreciate your investment in the current application thereby deferring the value created by the new application? Hopefully my point is clear: Planning could have avoided a very large problem. In an emerging technology, space standards are being hashed out and initially there will be numerous competitive products. Having a strategy to attack the problem will allow you to get the largest bang for the buck and avoid costly repurchases among other benefits.

Of course, you can't wait forever before beginning your wireless activity. A strategy simply enables you to get started in a more logical and directed fashion. Your wireless strategy will not ensure that every infrastructure investment is perfect and reusable, but it will ensure that you don't miss easy cases where a common technology can be deployed to support multiple applications. The 80/20 rule clearly applies here: The strategy process will ensure that you know 80 percent of eventual wireless applications, while 20 percent will change or emerge as new requirements down the road.

3. Reduce Cost/Increase Revenue

Ultimately, the argument for developing a wireless strategy rests with the potential for creating value for the business. It is our firm belief that taking the time and effort to develop a mobility strategy for your enterprise will ultimately allow you to create significant tangible business value. This value can be created in numerous ways:

- Identify more areas of the business where wireless technology can have a positive financial impact. For example, in wireless Help Desk or maintenance applications it becomes easy to create an ROI analysis. There is considerable cost savings from having issue or trouble ticket information with the employee at the site of the problem.

- Avoid making investments in places where wireless technology can't produce positive financial returns. Deploying wireless devices with general-purpose Web browsers without any custom applications usually finds employees surfing for information that isn't necessarily relevant to their job. The appropriate business applications must come with the device.

- Execute projects in the most logical fashion. For example, deploy solutions with the highest potential for a return. Defer projects if the technology infrastructure needed to deploy it is not yet ready for prime time.

As described in Chapter 1, there are numerous places where wireless enterprise solutions can create tangible business value for the enterprise. These will vary dramatically from company to company, which makes the strategy development exercise that much more important.

Who Needs One?

Should every enterprise develop a wireless strategy today? What types of businesses should consider this exercise? Let's start with a short history lesson on what businesses have already implemented fairly comprehensive wireless strategies and move to the more critical discussion of determining if this process is right for you.

The Wireless Enterprise Leaders

Although wireless and mobility technologies are still fairly new on the scene, several firms have been successfully leveraging these technologies in their operations for years. These companies have deployed wireless as a competitive weapon in their business. Not surprisingly, the companies that have been quick to adopt wireless technology have been major players in their industries and have had sizable mobile workforces. Consider the cases of FedEx and United Parcel Service (UPS), two of the dominant package handling and distribution companies in the world. Both FedEx and UPS have been pioneers in deploying wireless solutions. As early as 1990, UPS had developed and deployed a wireless solution for its thousands of brown parcel trucks termed Delivery Information Acquisition Device (DIAD). This enabled drivers to communicate through cellular networks directly to UPS' back end systems. Today, UPS has continued its efforts in wireless technology by deploying package tracking statistics and other information to their customer base via a myriad of wireless technologies.

FedEx has taken a similarly aggressive approach in adoption of wireless technology. When the wireless infrastructure in place in the United States was not sufficient to meet the company's requirements for capability, reliability, and coverage, FedEx developed its own network at a very significant cost to themselves, and today this network (termed the Digital Assisted Dispatch System) has become one of the world's largest private radio networks. Although costly, the network build out and corresponding development of mobile applications gave the company's mobile workforce the ability to deliver more packages per worker at faster speeds and with greater reliability. The company's follow-up study determined that productivity leapt by 30 percent following the introduction of wireless technology to its field force applications. This was necessary to back up the company's claims for guaranteed delivery times. The deployment of mobility technology has given FedEx the ability to track the location of every package in its network. Today this information is pushed to customers via the Internet and wireless devices, a huge competitive advantage for FedEx in the battle for business from online consumers and employees.

Frito-Lay also leads the way in the field of wireless and handheld technologies. The company's mobile workforce is considered pioneers in their use of technology, as they use this to effectively manage inventory levels at the hundreds of thousands of locations where the company stocks its product. Effective inventory management ensures low investments in finished goods inventory (and, hence, lower working capital required) while ensuring that stockouts are minimized. This maximizes revenue.

Many of the pioneers in wireless technology deployments are either transportation and logistics companies (T&L), or others for whom T&L is a significant component of their overall business. Figure 2.2 depicts the trends anticipated in the adoption of wireless technology by industry sector. This chart will perhaps give you some idea of whether or not your company is a good candidate for wireless technology adoption. If your industry has already adopted wireless or is high on the list for adoption, you are likely a good candidate for a wireless strategy. Later in this chapter we will go into more detail on how you can determine if your company would benefit from development of a wireless enterprise strategy. If you answer yes to the following questions, you should definitely explore the idea further:

- Does my company employ a high percentage of mobile workers?
- Are my customers mobile?

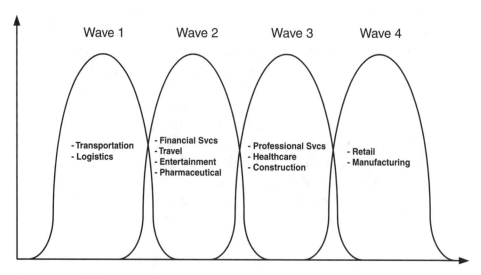

Figure 2.2 Wireless adoption waves.

- Do my customers demand information and transactions based on time or location sensitivity?
- Have my competitors adopted wireless technology for their business?
- Have my customers or suppliers adopted wireless technology for their business?
- Does my company have a successful track record of implementing new technologies?

If you are uncertain as to whether or not your company could benefit from wireless technology, fear not, for one of the key components of the strategy exercise is determining the potential to your company of going wireless. If in doubt, you may want to complete just the early phases of the strategy exercise before committing to the remainder, so you can be sure of the value the strategy process will bring to your company.

Components of an Effective Wireless Strategy

Hopefully, at this point we have you convinced of the need to complete a wireless strategy for your enterprise, assuming you fall into the high potential group. Earlier I described the need to take a broad definition of the term *strategy*. This meant we would include planning the rollout of the strategy within our definition. What do we really mean when we say wireless strategy? What are the components of a successful strategy? For our purposes, we will organize the components into the four main impact areas. In each of these impact areas, wireless technology will be felt to a smaller or larger degree depending on your company and the industry in which you operate. The four main impact areas are

1. Business model
2. Revenue enhancement
3. Cost and productivity
4. Information systems

Business Model Impact

It is popular in the press today to point to Internet businesses as a craze or a fad whose time has passed. After all, the strongest Internet companies remaining are traditional brick and mortar players like Wal-Mart who have been in business forever. The Internet startups, from business-to-consumer (B2C) companies to business-to-business (B2B) powerhouses are a skeleton of their former selves, if they exist at all. Yet few of the CEOs at America's leading companies would disagree with the statement that the Internet has profoundly changed their company and their industry. Consider companies such as General Electric, one of the oldest and most successful companies in business today. You might be surprised to know that they operate one of the largest Internet-based businesses in existence today (largely through their online Web business GE Global eXchange Services). The Internet as a disruptive technology has made a huge and positive impact on the business model of GE.

Wireless and mobile computing technologies have the potential to create a similar impact on business models in the future. If we think of a typical business model as a blend of

- Products and services
- Customers
- Delivery channels
- Supply chain

we can better understand the potential wireless technology impact.

Products and Services

For some companies, wireless technology will change their product or service offering. This is perhaps the most direct and far-reaching impact a technology can have on a business model, changing the company at its very core. As such, this type of change is typically reserved for companies operating in the technology space. In the wireless world, companies such as PumaTech and Synchrologic have rapidly modified their product offering to take advantage of the rapid proliferation of handheld computers from Palm, Microsoft, and others. Sprint, MCI Worldcom, and other telecommunications companies are rapidly creating new service offerings in the wireless data space to capture the growing demand of their customers for wireless data transmission capability.

phone calls during the day. Asset costs (as measured by a percentage of revenue) fall while the company meets its growing customer demand.

Customer support costs. Perhaps your company has a customer group that is highly mobile and demands information about their order status, delivery status, account balance, or whatever, wherever they are. By building mobility solutions for these customers so that they can access this information when they need it, self-support rises and your need for expensive (and maybe even around the clock) support declines. This is clearly on the minds of those in the financial services industry, many of whom are working feverishly today to build and deploy wireless applications for their customers. By the way, these support costs don't just apply to your customers; they could apply equally to your mobile workers and to your company's suppliers. I expect that these types of mobile customer support solutions will be commonplace quickly in industries such as package shipping and business travel management, where solid Web-based customer support solutions are already in place.

Customer acquisition costs. Customers . . . who needs them? My guess is your business does. For most companies, obtaining new customers can be an expensive proposition. Wireless technology may help your company in two ways. First, wireless solutions can serve to increase customer satisfaction by giving customers the information they need about your company when and where they need it, whether this is provided directly to the customer or through one of your mobile employees. A happy customer is less likely to go to a competitor, lowering customer turnover and minimizing the need to find new customers to replace those you have lost.

Second, some companies may be able to reduce their costs of marketing to potential new customers (and hence lower their acquisition costs) by using mobility technology. Consider location-based services, which can enable a company to target a customer when they are in a location where they are most apt to buy your product or service. For example, a consumer might voluntarily enter information in their user profile indicating their fondness for oriental cuisine. When they come within range of your restaurant, a service notifies them of your menu and pricing (either passively or actively). Voila, you have a new customer! Although this example is somewhat crude, I believe it does make the point about the power of mobile technology in customer acquisition. The potential improvement in target marketing should serve to lower customer acquisition costs if done correctly.

Don't forget that the cost savings potential of wireless technology will have to be balanced against the incremental costs of a wireless implementation, including IT infrastructure, process redesign, application development, and the like. These costs will be covered more fully in a later chapter.

Technology Impact

The final impact area we will consider relates to the technology itself. If you happen to be employed by your company's IS organization, you might already be starting to quiver at the number of potential wireless devices and applications and the impact this will have on your department in the coming years. The good news: Wireless technology vendors are considering the needs of the enterprise very carefully as they develop their devices, applications, middleware, and networks. The bad news: We are still in a very early stage of wireless technology development, with hundreds of potential vendors to choose from and no definite picture on the final standards. Unfortunately, due to the increased rate at which new technologies are being adopted as discussed in Chapter 1, companies can no longer wait years before adopting new technologies, or else they will risk a significant competitive disadvantage.

Wireless deployments can be a logical and effective component of your company's overall technology plan or a major fiasco that saps your resources and produces little or no benefit. Two simple rules will help to ensure that your company avoids the fiasco outcome. The first technology rule is to be sure to link your wireless strategy to your overall IT strategic plan. If you don't have a company-wide IT strategic plan, you should probably start with one before even thinking about your first wireless project. Having a solid overall technology strategy and plan to which your wireless strategy can be linked is important to successful outcomes in your wireless deployment. For example, if you are negotiating with your telecommunications carriers for a new global contract for your company, be sure to include your wireless data requirements as a part of this negotiation rather than as a separate exercise. Don't plan your IT Help Desk needs without considering the need to support new mobile devices and applications.

The second technology rule is to ensure that the development of your wireless strategy considers the technology impact of your desired wireless initiatives. Just as your wireless strategy would be incomplete without careful consideration of its impact on revenue generation and productivity, it likewise would be incomplete if the strategy did not carefully evaluate specific wireless technologies. A few examples will help to bring this concept to life:

- How do I leverage my wireless infrastructure from one project to the next, including devices, connectivity methods, and programming languages?

- Which wireless technologies are most appropriate for my organization based on the current skill sets of my people?

- How scalable is the synchronization middleware we intend to use?

- How will I manage upgrades to software applications where the code is resident on various handheld devices?

Hopefully these examples provide the flavor of the types of questions you will need to consider when evaluating the technology impacts of your wireless plan. Future chapters will describe specific wireless technologies in more detail, so you will better understand the most critical questions to answer for your organization.

The Wireless Strategy Development Process

"Great," you may say. "Now I understand what a wireless strategy is and why I need one. How do I go about creating a strategy for my company?" The first thing to understand is that developing a wireless strategy is a process. If you follow the process logically and deploy the correct resources against the process, you should achieve an effective outcome. In this section we will introduce the major components of the strategy development process at a high level. We will conclude with a discussion of how to staff your strategy team and the role of prototypes in the process.

Figure 2.3 presents a graphical overview of the approach or methodology we will be discussing in this section. Note that the order in which the strategy development process is completed is critical, as each set of activities build on work done in the previous section.

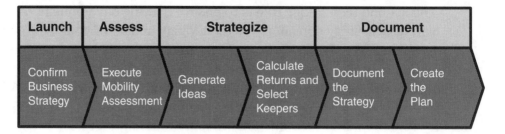

Launch	Assess	Strategize		Document	
Confirm Business Strategy	Execute Mobility Assessment	Generate Ideas	Calculate Returns and Select Keepers	Document the Strategy	Create the Plan

Figure 2.3   Strategy development methodology.

Let's proceed into our review of the methodology by tackling the first step in the process.

Confirm Business Strategy

Before we head too far in the wireless direction, it is important that the project team have a clear understanding of the company's overall business and IT strategy. If your company's business strategy is not clearly defined and understood by the team, it will be extremely difficult to capture what I will call the big impact ideas or those that have the greatest potential for driving economic value for the organization. Often a strategic plan may be available, but is somewhat dated (say, more than a year old). The team may be able to update the business strategy rapidly through a series of executive interviews after a thorough review of the plan. Other times a more comprehensive business strategy development process is required before proceeding. Don't think you can get out of this set of activities if your company happens to be a nonprofit hospital, school district, state government, or other organization. These companies also need a clear strategy for the organization, which ties directly to the mission or vision statement.

Execute a Mobility Assessment

Now the fun begins. Once the team has developed a comfortable understanding of the business strategy, they need to begin work on a mobility assessment. A mobility assessment is a preliminary evaluation of places where mobility is important in your operations. This assessment sets the stage for the identification of real opportunities for the deployment of wireless technology. In consulting parlance, you might consider the mobility assessment akin to a current state evaluation or as-is analysis. The team needs to identify opportunities where mobility is (or should be) an important part of your operations and rate how your company is performing in these areas. Let's discuss several factors you should consider in your mobility assessment to help make your job easier.

Time and location sensitivity. Not to sound like a broken record, but the first thing to consider in the mobility assessment is processes where time and location sensitivity is high. It is not always obvious where this is the case. Look for places where cycle times are long, and you could find a time sensitivity that is not being met. In many cases, a process with a long cycle time exists because of a highly inefficient process, which could be caused in part by a lack of good information about that process. Look for processes where cycle times are short, and you are likely to discover a time sensitivity that is being met, although perhaps not in the most efficient manner. To discover location sensitivities, look for instances where your workers, customers, or suppliers work in multiple locations during the course of the week and where information needs are high.

Number of mobile workers, customers, and suppliers. This may be the most obvious component of your mobility assessment, but it should not be overlooked. Do you have a large number of mobile workers? Remember that this isn't just the salesperson on the road five days a week. It can also be the executive who travels between all of your company's locations, the lab technician who moves throughout a large research office, the warehouse worker on the move picking and packing shipments, or the shop floor quality inspector moving about the plant evaluating production quality. Let's not forget about your suppliers and customers. Do your customers work in a highly mobile fashion, such as is the case for the pharmaceutical manufacturer whose customers include physicians? How about your suppliers, such as the pulp and paper mill industry, whose suppliers are in the woods chopping down your company's raw materials? In general, the higher your company's percentage of mobile workers, customers, and suppliers, the higher the value potential of wireless technology.

Missing opportunities to serve customers. This is a bit trickier to assess, but nonetheless important. Considering your company's products or services, it is useful to understand if there are potential customer groups who are underserved because of their mobile nature. For example, consider the online stock trading company. If they serve a highly mobile group of customers, it is possible that they are missing out on trade activity if they don't fill the need for their mobile customers to do business wherever they travel. Missing opportunities like these can mean lost revenue or even lost customers to competition that delivers mobile applications.

Holes in the information chain. Accurate and timely information is the lifeblood of most companies these days. More timely and accurate information on the production floor keeps automotive assembly plants humming along and avoids costly downtime that was the norm not so many years ago. Think of your business as a series of information chains, which accompany each of your core business processes (see Figure 2.4). This figure depicts a typical set of business processes for a manufacturing company. Each component of the process represented includes a corresponding set of information requirements for that process to work successfully. If you have holes in any of your information chains, your business processes may be operating in a sub-optimal manner. These holes can be caused by either gaps, where information is just not available at a key point in the process, or time delays, where information is available but not in concert with the execution of that step in the process.

Consider the case of the apparel manufacturer for teens who depends on timely and accurate market research to ensure that the company is

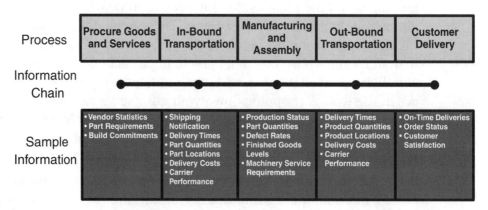

Figure 2.4 Information chains.

in line with rapidly changing demands. This company could have an information hole due to a time delay, as critical field research is reported long after it is collected. Or consider the manufacturing company that depends on thousands of parts and services provided by its suppliers. This company may know everything about inventory once it is in their plant, but what about when it is still several steps back in the process, either at a supplier's location or in transit? This is an example of a hole due to a gap in the company's information chain.

Be thorough in your examination of the organization as you conduct the mobility assessment. Use organization charts, process diagrams, and whatever else may be available to ensure that you cover the broadest spectrum possible of operations inside (and outside) the company to get an accurate picture.

Generate Ideas

Once you have your mobility assessment complete, you should have a pretty good understanding of where mobility stands in importance to your organization and how you are performing today. As you conduct the assessment, be sure to look where technology is already being deployed, for this will be an obvious first place to examine to see if wireless technology can improve the status quo.

Now it is time to turn to the fun part of the process, developing the set of new mobility opportunities. You may already have a pretty good set of ideas in mind from things that were very clear areas of improvement from the mobility assessment. You can add to your plate of ideas by implementing a few other techniques. Remember, in this phase you want to develop the broadest set of ideas possible for improving your business through wireless technology. This means considering areas where the technology may frankly not be ready to support your ideas. Don't worry about this fact right now, as we will come back to it at a later time. Let's turn to the techniques we can deploy in this phase of the project.

Best Practices

The idea of best practices is not new, but it has been considered more thoughtfully in the past 10 or 20 years than anytime before that. This is a bit tricky to define, but for our purposes we will consider best practices the idea of generating new ways of doing business by employing what has worked successfully for others. Many companies generate best practice ideas through formal benchmarking exercises. In these cases, companies will reach agreement with several other organizations to benchmark one or more business processes. The companies can all come from within the same industry or from companies across multiple industries that share common business processes.

This exercise can generate extremely detailed information for the participants, but can also take a significant amount of time. You might want to consider employing the aid of an outside firm with expertise in wireless technology and how it is applied to business processes and business strategy. These firms have the ability to draw on their entire organization and collective experience of working with many different organizations, which they can then apply to your business. Also, don't fail to consider generally available information sources, such as business periodicals like *Fortune* or *Business Week*, or specialty magazines such as *mBusiness*. You can often draw many best practices from these periodicals.

Competitive Analysis

It is often said that imitation is the greatest form of flattery. This applies perhaps most aptly in the business world, where companies are quick to replicate their competitors' best actions lest they gain some advantage in the market. This can be a great mechanism to identify ideas for mobility-based technology. A careful evaluation of what your competitors are doing may turn up some solid ideas for your company. Don't restrict yourself to competitors in your local geography. Consider what your competition is doing around the globe. In the wireless world, the United States is not considered the leader, at least according to several important metrics including wireless data infrastructure, wireless coverage, and subscriber rates. Consider places like Japan or Finland for ideas, where wireless penetration is extremely high and a common infrastructure provides a better foundation for more advanced deployment of wireless technology.

Internal Generation

Don't forget to carefully consider and capture all of the ideas you can from within your organization. You probably started this exercise when you conducted your mobility assessment as you interviewed key business leaders. Continue this process during the idea generation stage, and go back and cover not just the business leaders, but also the mobile workers and others for their ideas. You might also want to spend some time interviewing your customers and suppliers, as they are often a very valuable source of information and are not afraid to tell you what they would like to see different in working with you.

Once you have a rich plate of ideas, spend some time to organize the ideas. You might want to sort them according to where they impact the business or perhaps according to the type of technology necessary to execute the ideas. Although this is not the only grouping of ideas we will do, and may not be the last, it will make the later phase activities much easier.

Calculate Returns and Select Keepers

Now that we have determined our ideas and spent some effort to logically organize them, we need to figure out what each idea (or group of ideas) is worth. This is perhaps the greatest challenge in the entire strategy development process. Your company's CFO and CIO will be hovering over your every move in this activity, as it is critical to accurately calculate both the costs and benefits of each strategic recommendation made. The reality is that during strategy development, you are unlikely to drive cost and benefit details to an exacting level of detail. This is okay, for early in the process we are more interested in order of magnitude than exactness. In the following section, we highlight several steps to follow in this activity and some pointers to get you going.

Calculate the Benefit

In the previous step you spent time generating ideas where mobility technology could create value. It is important to keep an open perspective during this activity so as not to restrict creative thinking. Now, however, it is time to get a bit more precise on the potential economic value from each of these ideas. The level of detail and precision of these measurements will depend in large part on the culture of your organization. For some organizations, a simple high, medium, and low benefit estimate will suffice, although you should still establish ranges of potential economic value (high equals greater than 10 million dollars, medium equals one to 10 million dollars, and so on). For other organizations, financial calculations will be necessary to satisfy the top brass. What if one of your idea's largest benefit is greater customer satisfaction? Can you put a financial value to this? I believe you absolutely can quantify something like customer satisfaction and that it has a real financial impact for most businesses. For example, customer satisfaction often translates into reduced customer turnover, which equates to savings in customer acquisition costs. Or it may translate into higher average customer spending per transaction, which will boost your top line. Work through the benefits of each idea independently or in groups where ideas are interrelated. If you are not sure about grouping ideas, evaluate them independently and group them later during the planning phase of the process.

Figure Your Costs

Once the benefits are understood and quantified, move quickly into the cost estimation stage. We dedicate an entire chapter of this book to calculating costs for wireless projects, so we will limit the detail here. Remember that you must calculate one-time costs, such as up front software licenses, infrastructure build out, and consultant implementation fees, as well as any ongoing costs, such as software maintenance fees, airtime access, and specialized IT support. Don't overlook other important cost components associated with any business process change or introduction of new technology, such as training. Finally, recognize that your company may actually experience a dip in productivity during the rollout of the new systems, as people spend time learning how to use the new technology.

What is the best way to evaluate costs that can be shared across multiple wireless initiatives? Consider for example, an investment in a wireless LAN for your warehouse and plant floor. The investment could be used to support a wireless pick and pack inventory management solution, as well as a new wireless system for managing plant floor machinery uptime. This same LAN may be used to support other wireless solutions that are not yet identified. Although there is no easy answer to this problem, consider the following tips:

- Don't assume too many future initiatives when you spread the costs for a fixed investment, as you will likely replace most of the infrastructure (wireless LANs, handheld devices, and the like) after a few years when more advanced technology or new standards arrive.

- It is acceptable to group initiatives together when it makes sense to implement these new ideas as part of a larger project. For example, the two previous solutions might logically be implemented as a single project to wirelessly enable your factory. This enables you to share some project costs that you would otherwise count twice, such as project management costs, plant floor downtime costs, and the infrastructure itself.

- Be careful of splitting infrastructure costs across multiple initiatives. Although it may seem intuitive to allocate your costs for your wireless LAN to several initiatives, if only one or two initiatives get done, you will have underestimated your costs. This may not seem fair, but if one of the projects you are counting on is scheduled to start three years in the future, there is a real chance this project may never take place. For example, changes in economic conditions or company leadership have been known to cause the deferral or cancellation of projects at many companies.

Review Technology Impact

Now that we have generated our ideas and calculated the potential returns, we should revisit our company's technology strategic plan to evaluate how our wireless ideas fit within this larger plan. Although some people may argue for doing this prior to generating ideas, we would suggest that this might limit the creativity of your thoughts if you feel constrained by the current or planned IT architecture of your organization. If we time our review of the technology impact of our wireless initiatives after we have generated our ideas, we will still gather crucial information to help in calculating the costs of implementing our ideas. We should also be able to gather useful information about topics, such as development languages, preferred vendors, and planned technology infrastructure, that will be useful in calculating our wireless project returns and in generating our master implementation plan.

Determine Return on Investment

Once you have your costs and benefits clearly understood, it is time to calculate your returns. As expected, there are numerous ways to calculate your returns, including the following popular choices:

- Return on Investment (ROI)
- Return on Assets (ROA)
- Economic Value Add (EVA)
- Payback period
- Net Present Value (NPV)

Most of these methods entail calculating the costs and benefits over units of time. For example, you might calculate costs and benefits by year for a five-year span. These methods are a bit of both art and science, so do your homework and be sure to learn what methods are most acceptable at your company before you begin. Companies tend to be very particular about how ROI is calculated in part because they may compare the return of doing your wireless project against other potential projects, and this is much easier if the calculation method used is the same for each project.

The previous methods of return can involve weeks of effort in some cases to accurately calculate returns. Many companies will want a much shorter process for estimating returns in the strategy development phase. Sometimes it is acceptable to calculate a very high level estimate for both costs and benefits. This can help in the next phase to screen out low return ideas and get focused on those with the greatest potential for payback to the organization. Figure 2.5 highlights the results from such an ROI calculation. In this instance, the strategy team was focused on quickly identifying ideas with high probability payback while eliminating the losers (or at least eliminating those ideas whose time has not yet come).

Document the Strategy

I hope that throughout this process you have been doing a careful job of documenting your analysis and your findings. My guess is that you would have a hard time keeping all of this information straight if you

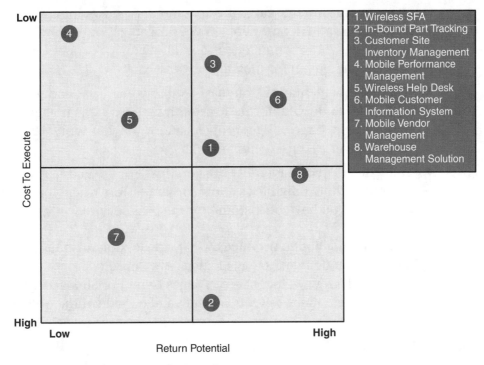

Figure 2.5 Wireless cost and return chart.

weren't documenting. Documenting your wireless strategy should be the easiest step in the process, but it is also the easiest step to skip or do poorly. Don't underestimate the importance of doing a thorough job of strategy documentation. Although some people may accuse you of spending effort to create *shelfware* (the derogatory term used to refer to sometimes brilliant pieces of business strategy documentation that never leave the CEO's bookshelf), you will regret not carefully documenting your strategy. Consider a few important reasons for documentation:

- You may quickly forget key elements of the strategy development process, and a complete documentation set will serve as a great refresher.

- A well-written strategy document is one of the best ways to quickly communicate your ideas and recommendations throughout a large organization.

- Having your findings carefully documented allows for a review with all divisions and departments in your organization who will be affected by your recommendations, allowing for proper input to the process and buy-in for moving forward.

- During implementation you may need to come back and present your ideas and ROI information multiple times. This is especially true for long, large budget projects and when your organization has had turnover at the executive ranks.

- You will spend considerable time on this process and may need to have something tangible to show for your effort when your boss asks what you have been doing for the past several weeks.

Don't feel compelled to produce a 400-page document suitable for publishing, but you do need to have logical, organized documentation of the work and a few well-conceived pages of text or slides to quickly present the main ideas to groups inside and outside of the organization.

Create the Plan

The final step in the strategy development process is to create the implementation plan (and if you think you don't have to document that plan as well as the complete strategy, think again). Your plan is the logical transposition of the strategy into action. Without a well-defined and executable plan, your strategy truly is meaningless. First, let's tackle the components of the plan. There are nearly as many different types of workplans as there are strategies for investing in the stock market, and this is not a book on how to create effective workplans. Despite the differences, each workplan will typically include the following components:

Action steps. Every plan will lay out in some detail the major action steps that need to be taken to complete the initiatives. For a broad scope plan covering several initiatives, it is best to have high-level action steps with additional detailed action steps following for the early-stage activities. The action steps might simply include the name of each initiative and nothing more for a summary view of the plan.

Timing. Even the highest-level workplan needs to include a description of the time it will take to accomplish the various activities in the plan. Although detailed workplans for a specific project will define the timing in increments of days or even hours, plans covering an entire wireless strategic implementation are better represented in weeks, months, or perhaps in some cases, years.

Resourcing. What effort will it take to accomplish the tasks at hand? Who will accomplish the work? These critical questions are answered by considering project resourcing. Detailed plans will take resourcing down to individual project team names, while summary plans typically include only a measure of the number of full-time equivalents (FTEs) required.

Dependencies. The plan dependencies will show which steps need to occur prior to others taking place. For a wireless strategy workplan, the dependencies might be to simply show which sub-projects need to be completed before other projects can begin. In the wireless world, dependencies can be crucial, as one project's infrastructure may be a prerequisite for another project.

The following chart attempts to lay out a fairly typical summary level workplan, representing the cumulative plans of the entire wireless strategy (see Figure 2.6). It may or may not make sense for your company to create a master wireless workplan. You may find it more efficient to make IT workplans by division or geographic region with the wireless projects interspersed.

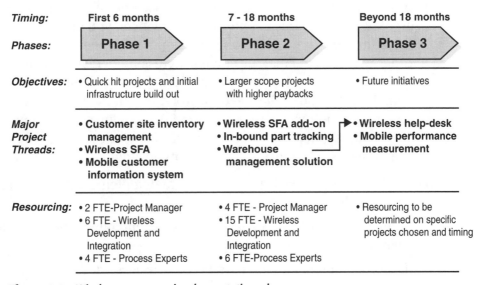

Figure 2.6 Wireless summary implementation plan.

A few final thoughts on developing wireless project plans

1. Do try to plan high ROI projects early in the implementation cycle, along with short, quick payback projects. These projects will set a good tone for wireless at your company and build momentum to complete projects with less resounding ROIs or longer implementation cycles.

2. Group smaller projects together if by doing so you can save time and cost in implementation, as long as ideas fit together logically by leveraging required infrastructure, and so on.

3. Don't throw out expensive ideas with low or negative ROI. Although this may seem crazy, remember that wireless is very much an emerging technology, and as such, ideas that look too costly or infeasible to implement today could become inexpensive in the future as the technology advances and prices decline.

4. Because many wireless solutions rely on data from multiple back end systems, don't underestimate the effort to get data out of each system and integrated into the overall solution.

What about Prototypes?

For many people, the idea of a rigorous strategy development process may seem outdated. They argue that it is better just to get started on your first project and see where this leads. Hopefully the arguments presented earlier in this chapter have convinced you of the benefit for completing some type of strategy development process. Yet there is another way to begin making some progress immediately with wireless technology, even during the strategy development phase. This can be accomplished through the use of prototyping. In the wireless world, prototyping is defined as the process of developing a small, limited functionality application or deployment of a wireless technology to a subset of users. The purpose of a prototype includes

- Getting your hands dirty with the actual implementation of a specific wireless technology

- Achieving meaningful use of wireless technology without making a significant investment

- Building interest and momentum for wireless solutions in your organization

As an example, look at Jeff Hawkins, inventor of the Palm platform. As he was considering the qualities of a successful handheld device, he understood the need to keep its size and weight to a minimum. To that end, Jeff carried a block of wood in his hand and shirt pocket that had the dimensions of the first Palm 1000 device. His prototype, although nonfunctional, served its purpose of giving him understanding about the size and weight of a successful handheld device.

Prototyping can also be done as a parallel exercise to strategy development if you have a deep enough team to accomplish both simultaneously. Prototyping can be a very powerful tool for your organization. You may want to develop prototypes as a part of each major project or phase of activity. It is also a great way to get early input into some of your ideas to see how they play in the real world. In Chapter 8, "Handheld Application Design," we cover many tools for quickly developing prototypes.

Getting the Right Resources

What are the best resources to deploy against your wireless strategy process? This again is highly dependent on your organization. If you tend to do all of your strategy and technology work in-house, you are more likely to want to do your wireless work with internal resources. If you tend to use outside companies and contractors, this approach will also fit your needs with wireless. Keep in mind that wireless is very much an emerging technology, and as such there are some great reasons to get outside help:

- The technology you deploy today will have a significant impact on your IT costs over an extended period of time, and outside specialists are likely to be more familiar with what technology is right for you.

- Wireless specialists may be able to bring some viable best practices from deploying wireless technology from their other clients.

- A good wireless consulting firm can help you build strong wireless development and implementation skills in-house so you can control more of this important technology in the future.

- You may want to supplement your own team with a small number of outside specialist resources where you have a few key skills missing.

Whatever your decision, remember that there are many firms anxious to compete for your business these days, and this is especially true in the wireless world. You can learn a significant amount from simply meeting with a number of software and services companies without committing to a decision to use their technology or services. Read this book. Invite them in and quiz them. Do they push one technology, platform, or data-delivery technique? Can they handle strategy and implementation? Are they skilled in knowing when to use one platform over another?

There will also be many companies that have the capability to tackle wireless projects with in-house resources, and in some cases they may not realize it. If you have skilled Java and C++ developers, these resources can often be utilized for development of native handheld applications. Web developers can transfer their expertise to develop solutions for the wireless web. Most wireless projects require integration to new or existing databases in SQL Server, Oracle, or DB2, and your skilled experts in these areas could be put to good use as well. Many times experts in data replication will have ideas on how to handle the business logic of data synchronization. Of course, you typically must be willing to sacrifice some speed in delivery of the final application as time will need to be spent on training for these resources, some of which will occur during the design and development process.

The Bridge from Here to There

Creating and documenting a corporate strategy can take significant time. Even if you decide not to go through the entire process, some level of thought must go into the plan before you begin to execute lest you waste dollars on infrastructure and cost overruns and have unsatisfied user needs. Once you've got it in place, executing this plan will be easier and well supported by the stakeholders.

Although strategy is key, sometimes it's hard to envision places where wireless will make a difference to your business. In the next chapter we examine some real world examples of how wireless technology can be used in different vertical markets and with consumer products and services. This may help jump-start your thinking on opportunities for this technology within your organization.

CHAPTER 3

Bringing Wireless to Life

As we discussed in previous chapters, wireless has the potential to be a disruptive technology that requires organizations to rethink the principles that drive decisions regarding their business. This potential for radical change requires organizations to embrace this new technology and develop a well-constructed strategy that is congruent with the overall corporate value proposition and strategic objectives. Once this strategy is articulated, initiatives should be deployed to increase corporate performance primarily along three dimensions: increase revenues, reduce or avoid costs, and improve competitive positioning. These initiatives will help solve fundamental problems that have plagued businesses for years by addressing time and location sensitivity issues of underlying business processes and activities (see Table 3.1).

Like most technology initiatives, maximizing value by implementing wireless solutions usually requires changes to people (capabilities and organizational structure), business processes, and technology. These changes may be substantial or fairly minor and will depend on the current state of the organization as well as which processes are being wirelessly enabled.

Table 3.1 Problem Areas Wireless Solutions Help Solve

AREA	PROBLEM SOLVED
Decision support	Decisions are delayed due to a lack of information mobility. When a situation arises, having access to the required data can shorten cycle times and save costs.
Performance support	Many decisions on critical issues are made with out-dated information sacrificing effectiveness and performance.
Efficiency	Mobile workers increasingly perform redundant activities (data extraction and entry) due to a lack of access to back end systems while in the field.
Utilization	Avoiding asset downtime (both human and machine) by supplying and capturing certain information remotely to drive up productivity.
Customer satisfaction	Customers are increasingly becoming mobile and will demand access/service when and how it is convenient for them. Customer satisfaction will be driven by providing multiple convenient touch points to allow the customer to manage the relationship.

Whether your organization is a high-priced professional services company or in the manufacturing arena supplying commodity components, wireless technology has the capability to help differentiate your offering and improve your employees' productivity. As depicted in Figure 3.1, wireless technology can help most organizations across a number of aspects of their business.

In this chapter, we will bring to life examples of wireless technology that can help organizations drive substantial value. We present this chapter as a way to stimulate your thinking. We've found on many initial visits with clients that wireless seems to be an intriguing way to cut costs or increase revenue, but they can't envision places within their own organizations it can help. The topics we have chosen to highlight are by no means the only applications of this technology, rather they are real examples of applications that we have had experience building and delivering for clients or believe will be available or popular in the near future. The examples we will cover include

Wireless in Customer Relationship Management (CRM). Assisting
the buying process

Process:

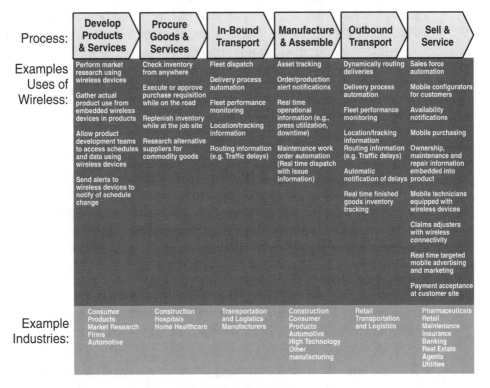

Develop Products & Services	Procure Goods & Services	In-Bound Transport	Manufacture & Assemble	Outbound Transport	Sell & Service
Examples Uses of Wireless:					
Perform market research using wireless devices	Check inventory from anywhere	Fleet dispatch	Asset tracking	Dynamically routing deliveries	Sales force automation
Gather actual product use from embedded wireless devices in products	Execute or approve purchase requisition while on the road	Delivery process automation	Order/production alert notifications	Delivery process automation	Mobile configurators for customers
Allow product development teams to access schedules and data using wireless devices	Replenish inventory while at the job site	Fleet performance monitoring	Real time operational information (e.g., press utilization, downtime)	Fleet performance monitoring	Availability notifications
Send alerts to wireless devices to notify of schedule change	Research alternative suppliers for commodity goods	Location/tracking information	Maintenance work order automation (Real time dispatch with issue information)	Location/tracking information	Mobile purchasing
		Routing information (e.g. Traffic delays)		Routing information (e.g. Traffic delays)	Ownership, maintenance and repair information embedded into product
				Automatic notification of delays	
				Real time finished goods inventory tracking	Mobile technicians equipped with wireless devices
					Claims adjusters with wireless connectivity
					Real time targeted mobile advertising and marketing
					Payment acceptance at customer site

Example Industries:

Develop Products & Services	Procure Goods & Services	In-Bound Transport	Manufacture & Assemble	Outbound Transport	Sell & Service
Consumer Products Market Research Firms Automotive	Construction Hospitals Home Healthcare	Transportation and Logistics Manufacturers	Construction Consumer Products Automotive High Technology Other manufacturing	Retail Transportation and Logistics	Pharmaceuticals Retail Maintenance Insurance Banking Real Estate Agents Utilities

Figure 3.1 Business processes affected by wireless.

The wireless executive. Corporate performance information on demand

Automating the mobile worker. Sales force example

Changing the value proposition of products. Rethinking what and how to sell to consumers

Wireless and CRM

The customer is king. The customer's perception is reality. Without customers you're dead. These statements although trite are absolutely true and seem to be at the top of the mind for most executives in business today. Although ignoring and mistreating customers has never been a particularly good business strategy, the speed and severity of the punishment for not pleasing customers has never been greater. Customers today have more options than ever, and with information at their fingertips, the cost of switching has never been lower. Many dot-coms that did not handle customer service well felt the pain and some actually went out of business in part as a result of poor customer service.

In today's business world no other topic is receiving as much attention as Customer Relationship Management (CRM). Although on the surface CRM may seem like just another buzzword within business literature, it really represents a strategic shift for many organizations. At its core, CRM is a change to an organization's mindset from an internal (that is, our company is the center) to an external (that is, customers are the center) perspective. This change can and should have profound ramifications for every function within an organization. It elevates the relationship a business has with its customers to be one of the most important assets the organization possesses.

Business publications are filled with examples of organizations in search of the Holy Grail of improved customer satisfaction. Some organizations are spending tens of millions of dollars implementing advanced information technology systems focused on the dual purpose of consistently exceeding customers buying and service expectations while providing additional revenue opportunities for the organization through more effective cross-selling and up-selling.

Wireless technology can play a very important role in your organization's CRM strategy. One of the critical elements of an effective CRM strategy is pleasing the customer with the buying process. Let's discuss specific examples of improvements that can be made to the buying process using wireless technology.

Revolutionizing the Buying Process

The buying process, whether a product or service, basically involves five stages:

1. Awareness/motivation
2. Research
3. Comparison/selection
4. Purchase
5. Fulfillment

Customers are becoming increasingly demanding. They are requiring that businesses tailor their process to match how they like to buy. Customers want to manage the relationship, and organizations need to be capable of helping buyers through all of these stages. As customers become more mobile, there is an obvious need for organizations to open the wireless channel for them.

The right wireless solution to help customers with the buying process will vary for each company and can be a fairly complex analysis. Wireless technology can be used by organizations to help customers across every stage of the buying process. Whether it is targeted mobile advertising to a PDA or using electronic micropayments for a candy bar from a vending machine, wireless technology promises to take the buying process to an entirely new level of sophistication. Not all applications of wireless technology are appropriate for every organization. As you would expect, the correct use of the technology depends upon the type of product being purchased and the degree of satisfaction with the current process. Product differences related to cost, complexity, brand image, and impulse versus planned purchase (among other characteristics) need to be considered to determine the correct application of wireless for the buying process. Let's take a look into the not-too-distant future at an example of how wireless can help a consumer with the buying process.

Wireless Purchasing

Abbey is a working mother of three with a very stressful job that always makes her feel as if she needs to stay in the office until the last possible minute before racing home to help her husband make dinner and get the kids ready for bed. Given her hectic schedule, she feels that superior organization of her life is a must. She has put in a process at home that allows each family member to enter items the house needs into the home PC that sits in the den. Items could include everything from Pop Tarts to toilet paper. Despite her best efforts, it always seems that she forgets to look at the shopping list and the family is consistently out of something, which requires her to run to the store for last minute items.

One day Abbey overheard one of her company's IT staffers discuss the latest handheld from Handspring that was equipped with Bluetooth technology and associated consumer friendly applications. Abbey is a gadget junkie at heart and loves new electronic toys. She knows that Bluetooth is a wireless technology that allows small form-factor devices to wirelessly exchange data without human intervention, and that many devices were now using it, including her home PC. Let's follow her through a typical day and see how wireless technology helps Abbey on a Saturday filled with errands:

9:00 A.M. Abbey is ready to go shopping and walks by the den on her way out the door. Her new Bluetooth-enabled handheld makes a quick wireless connection (without any interaction on her part) with the home PC. Her handheld now has the family's combined shopping lists.

10:00 A.M. On the way to the grocery store, Abbey walks past a sporting goods store and hears her PDA beep. The store (Bluetooth-enabled) is advertising tennis rackets at half off and the handheld alerts her that her shopping list indicates she needs a tennis racquet for her son's upcoming tennis lessons. Not knowing which to buy, Abbey uses her handheld to access the store's wireless network and do a quick product comparison. She narrows it down to two possible candidates. At that time the store's system knows that there is some flexibility in the pricing of one of the racquets. The store alerts her PDA that she can get one of the racquets at a 20-percent discount. Abbey selects that one, checks it off her list, and wirelessly authorizes the funds transfer from her handheld to the store from her handheld.

10:30 A.M. At the grocery store, Abbey uses her handheld (with a built-in barcode scanner) to scan grocery item barcodes and read nutrition and recipe information as well as receiving electronic coupons. After she is done, she uses the PDA again to beam her banking information to the store's cash register.

12:00 Noon. Next it's off to the local appliance store to look at new dishwashers. She really likes the new Whirlpool model and sees that the store is selling it for $459.00. Abbey uses her PDA to access the Internet and compare pricing. She finds it on sale, today only, at another store for $399.00 and decides to purchase it (all while staying at the original store). She uses her PDA to complete the transaction and schedule a convenient delivery time.

2:00 P.M. On the way home, Abbey stops to fill up with gas. She uses her handheld to transfer her credit information to the gas pump before pumping gas. The wireless network at the gas station detects Abbey's PDA using Bluetooth technology and sends a signal of what items are on sale inside. Because her handheld has two of those items listed, a flashlight and batteries, it is alerted that this station has them on sale. Abbey goes in, buys the items, and checks them off her list.

2:30 P.M. Abbey arrives home, walks in the door and past the den, and her PDA updates (automatically) the home PC shopping list and enters all of the purchases into the family's expense tracking software.

This hypothetical example shows the power of wireless to assist Abbey with her buying experience. We saw wireless technology assist Abbey in each stage of the buying process. Targeted mobile advertising helps create awareness, mobile Internet access assisted in research and comparison/selection, electronic payment helped with the final purchase, and fulfillment was aided by scheduling a convenient delivery time. Benefits exist across the buying process for this type of capability:

Targeted awareness/motivation. Organizations will increase revenues through personalized advertising and marketing, creating additional sales. In addition, effectiveness of advertising will be increased, enabling businesses to optimize their advertising spend. Consumers will feel like they are receiving pertinent information that relates to their individual needs and desires.

Streamlined research. The ability to search the Internet and perform product and vendor research whenever and wherever convenient increases a consumer's confidence and should decrease the sales cycle for a business.

Timely comparison/selection. Time savings for both the organization and the consumer due to the ability for a consumer to perform side-by-side comparisons of competitive products while in the store. Consumers will feel more comfortable with their purchases due to having the right information at the right time to make an informed decision.

Streamlined purchase. Electronically sending payment information using a PDA allows consumers to carry less cash and/or avoid the need for credit cards. This should speed the process and enable more consumption of goods from sources that traditionally only accept cash (such as vending machines). In addition, the ability to facilitate a purchase of products while physically standing in one store and buying from another will help consumers achieve the best price possible and extends the reach of many retail establishments.

Easy scheduling of fulfillment. Scheduling delivery from any location with a wireless device saves businesses time and effort and increases the satisfaction of consumers as they feel they are managing the relationship.

The Wireless Executive

An executive's job depends on having the right information at the right time to assess situations and help make decisions. Having visibility into current corporate performance is critical to consistently make correct decisions. Wall Street has become very impatient and intolerant of companies that miss deadlines and report surprises. Investors value management that is in control of the organization and swiftly takes actions on problems due to superior insight into daily operations. However, to truly have this capability requires access to information across the extended enterprise regardless of the time and location constraints.

The required information usually resides in disparate databases throughout the organization and historically has not been easy to gather, resulting in decisions based more on intuition and gut feel without supporting data. We would not suggest that anything can or should replace the intuition and gut feel of a good manager; it can, in fact, be a tremendous asset. However, for the majority of executives correct, up-to-date information is one of the most invaluable tools to help confirm this intuition and to make decisions. Furthermore, with the accelerated pace of change every industry is now experiencing, staying in touch with underlying facts is essential to adjust strategy and ensure execution continues to support overall corporate objectives.

As a result, many organizations have tried to implement executive decision support systems and/or balanced scorecard systems to provide information on critical metrics that drive value. Decision support systems enable executives to simulate a business environment and determine the correct course of action based on the assumptions defined within the model. Balanced scorecard systems display information on metrics that are pertinent to a given executive's sphere of influence and, correctly implemented, highlight the potential tradeoffs each executive must deal with on a daily basis. For example, balance to the measurement system of a retail organization will require that executives be measured both on inventory turns and customer fill rate. See Figure 3.2 for a screenshot of a wireless EIS system.

Taken together, decision support and balanced scorecard systems are designed to enable executives to actively manage value drivers and align behavior at all levels of the organization to ensure that strategic objectives can be realized. They typically query and report on back end data to allow managers to see relationships that previously were not visible. For example, let's say a U.S. manufacturer of roller blades is in need of increasing production for an upcoming spike in demand they are anticipating in Canada. A decision support system could enable them to see the impact across their balanced scorecard of various alternatives they may consider. Based on certain assumptions, they could see the impact of choices such as outsourcing a portion of production to a contract manufacturer, leveraging idle capacity in one of their plants in South America, and shipping the goods to Canada or running third shift in their U.S. plant and sending the goods north.

These tools work well in those instances where time and location are not a tremendous concern. However, many critical decisions are made away from the manager's desktop and possess high time sensitivity.

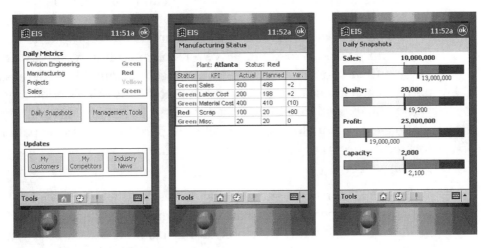

Figure 3.2 Handheld/wireless EIS system.

There are two problems with today's approach to provide managers with decision-enabling data:

1. Most systems do not support the mobile nature of a manager's job with easy-to-use, event-driven data that is integrated into an overall defined business process.

2. Few systems use exception-based reporting to flag early indicators and enable managers to take action before significant problems surface.

Wireless technology can help solve the remaining problems that exist with today's decision support systems. Let's look at an example of how wireless technology can help managers stay in touch with corporate performance and allow their organizations to respond more quickly to market movements.

A Day in the Life of a Corporate Executive

Alex is a busy executive in charge of an assembly plant making a popular vehicle brand at a large automotive company. He consistently finds himself on the go within the plant, corporate office locations, and at suppliers' facilities meeting with engineering, operations, and support departments responsible for ensuring the right vehicles are at the right location at the right time for consumers. Alex is keenly aware of the increasing competition from both foreign and domestic companies. He has a series of performance measures that are effective at focusing him on two primary objectives—happy consumers and a profitable company. These metrics foster creative tension and often require Alex to make difficult decisions that rely on correct information to assess alternatives and choose appropriate actions. Recently, Alex's company purchased all executives PocketPC (Windows CE-based handheld) devices that have a personalized executive dashboard that has the capability to access corporate information to stay in touch with the company's performance.

Today, Alex is in an off-site meeting discussing the new model changes to his vehicle line when his PocketPC beeps indicating a problem that needs to be dealt with. Alex looks at his dashboard and notices that production has turned to a red status and needs to be looked into. He taps the screen and sees that there is a problem with the Tennessee assembly facility and, specifically, that there was a defect found in a batch of cloth seats and consequently there are not enough cloth seats to satisfy today's production schedule. Alex immediately uses his handheld to check the estimated shortfall in seats and compare that to today's estimated delivery schedule. He sees that his seat supplier is due to send a shipment today; however, the shipment has already left the supplier and the quantity sent is still 200 short and does not allow him to meet today's schedule.

Alex's choices are to expedite another shipment of cloth seats to his plant to cover the shortfall or to reschedule production to vehicles with leather seats. Expedited shipments entail premium freight costs and are very much discouraged by Alex's company. In addition, the company does not like to build vehicles that do not have an assigned order (that is, they do not build to stock). Using his wireless PocketPC, Alex sees that the estimated cost of expediting shipment is $2,000. If possible, he would rather reschedule production to make cars with leather seats. However, this option depends on the current leather seat inventory level as well as committed orders in the schedule for the next few days. Alex accesses the inventory system, selects the leather seats, and sees that he has enough in inventory to satisfy the 200 cloth seat shortfall while avoiding the incurrence of an expedite charge. He then checks production schedules for leather seats for the next few days and sees that he has enough committed orders to build the 200 vehicles with leather seats and keep the plant from shutting down. Alex initiates the production change from his PDA and a crisis is avoided without having to spend an inordinate amount of time analyzing alternatives and searching for data.

Alex is still bothered by the defect found in the cloth seats so he uses his handheld to inquire about the circumstances. He notices that the defect has also caused an alert on his supplier quality metric and that this particular supplier has been performing poorly over the past two quarters. Alex e-mails a note to his supplier quality engineer to schedule a meeting tomorrow with the supplier to discuss remedies to the situation.

Most executives would agree that this type of solution could have tremendous benefit for their organization. However, we would be remiss not to mention that this type of integrated decision making depends not just on integration of the technology but on changing the mindset of managers as well as redesigning the core processes of the business. Unfortunately, the people and process changes are the most difficult changes to make within a company. However, we would submit that the change is worth the pain and those organizations that can implement this capability will have significant competitive advantage.

The benefits of deploying a wireless solution for executive information and decision support are based on providing real time information to enable better decision making more quickly. Therefore, the fundamental drivers of benefits are

Speed. Providing information regarding corporate performance anytime and anywhere decreases the decision-making cycle time, which can allow for time to market advantage.

Quality. Poor decisions are often caused by a lack of accurate data. Better information provided at critical times will reduce the number of bad decisions.

Benefit areas organizations can expect to influence as a result are

Revenue increase. Time to market advantages result in higher margin products and increased unit sales. In addition, shortened decision times on issues that directly affect customers should drive higher customer satisfaction.

Cost reduction. There can be a tremendous productivity enhancement of both executives and support personnel by streamlining decision making and having easier access to data. Process steps to solve problems can be eliminated, thereby reducing effort and costs.

Cost avoidance. Better information at the critical moment of decision making will enable organizations to avoid rework and the costs associated with incorrect decisions (such as, our premium freight example).

Automating the Mobile Worker

As we discuss several times throughout the book, time and location sensitivities drive the advantages for wireless technology. Time and location sensitivities are prevalent in virtually every industry and largely depend on the particular job function. Many organizations attempt to automate the mobile worker using a laptop PC; however the characteristics of a PC (primarily boot up time, size, and weight) make it a suboptimal solution. To be successful, any solution for the mobile worker must be an integral component of the overall business process and sufficiently help the worker to ensure adoption takes place. Solutions that are not successful tend to be those that are designed to make another person's job easier and offer limited benefits to the individual using the system in the field. For example, systems that serve the primary purpose of monitoring location and productivity of field technicians tend to be viewed as Big Brother types of solutions, and many field workers resent the associated implication. Also, complex solutions requiring a high level of intervention and effort that detract from time with a customer tend not to be successful.

A mobile worker that is prevalent in just about every industry is the sales representative. A salesperson's job is to represent his or her company to the outside world and successfully generate revenue by establishing solid relationships with potential and current customers. Outside sales people, those that predominantly sell by seeing customers (that is, not telemarketers), try to spend as much time as possible away from the office cultivating their territory and calling on accounts. A typical salesperson's job is becoming more difficult due to increases along three dimensions: customer expectations, competition, and the complexity of the products being sold. In many industries, customers are no longer buying just a product; they want a solution. For example, a major railroad company has realized that it needs to upgrade its sales representative capabilities to help clients with their transportation needs. In the past, transportation needs would have meant that the sales reps sold units of rail capacity, but now customers who are requesting railroads understand and assist with solutions to supply chain problems that may or may not include rail. The result is that most organizations are trying to adopt the consultative sale, requiring more information and account insight at the moment of truth, in front of the customer.

At a high level, what a salesperson does follows a fairly well-defined process. Each potential customer is uncovered as a result of a sales representative's prospecting activities. Once the lead is qualified and the salesperson is reasonably sure there is an opportunity to make a sale, he begins a different set of activities to sell the prospect on his goods or services. After the initial sale is made, he maintains the account and attempts to grow the relationship through selling other goods or services his company provides. As Figure 3.3 shows, there are many areas where wireless technology can be used to more efficiently and effectively manage the process.

The benefits of using wireless technology to help the sales process are best illustrated through an example. Let's look at a case study depicting how a sales rep can use wireless technology to help serve her client's needs while also maximizing the cross-sell possibilities for her firm.

A Day in the Life of a Sales Representative

The rail industry has been in existence for a long period of time and is still seen as a very viable and necessary mode of transportation within the United States. Like many other industries, organizations within the rail industry have found a need to add complementary services around the core rail transportation offering to differentiate their product. Katie is a sales rep at a major railroad company who is responsible for all retail establishments. Steve, an executive at Next Generation Electronics, her largest account, has asked Katie to come in on Friday and dis-

Process:	PROSPECT	SELL	MAINTAIN	GROW
Common Activities:	Identify Prospects	Multiple Phone Calls	Status	Lifetime Value Profiling
	Mass Market	Site Visit	Tracking	Cross-selling
	Answer Questions	Represent Marketing Programs and Discounts Available	Call Backs	Up-selling
	Manage Pipeline		Service Inquiries	Relationship Building
	Qualify Leads	Negotiate Pricing and Delivery Terms -Availability -Capacity -Pricing	Maintain Customer Data	Maintain Customer Data
	Record Prospect Information	Value Bundle		
	Produce Call Reports	Maintain Customer Data		
		Produce Sales Reports		
		Input Order and Customer Delivery Requirements		

Figure 3.3   Wireless helps the sales process.

cuss a recent bid Katie's company has submitted to handle Next Generation's West Coast transportation needs.

It's Friday afternoon and Katie has had a very busy day calling on clients. She is on her way to Steve's office and decides to use her wireless PDA to check for any relevant announcements and any recent service issues that her railroad has had with Next Generation. She accesses her company's back end systems to see that 99.2 percent of all shipments this year were on time and that they have received an exemplary status from Next Generation.

She also sees that Next Generation has announced a merger with Horizon Electronics. Horizon has been a target of Katie's for that last two years but she has been unsuccessful getting into the account due to a long-standing relationship between Horizon and Katie's number one competitor, RMB Express. Still using her wireless PDA, Katie looks into her company's competitor database and checks information on RMB Express. She realizes that RMB has a history of spotty performance and competes on price alone. In addition, they are not very strategic and have no known initiatives aimed at offering additional services for clients, even though outsourcing has gained in popularity recently.

Katie arrives at Steve's office. Let's listen to Katie's meeting with Steve.

Katie: Hi, Steve. How are you?

Steve: Doing well, but in a bit of a rush, Katie. I only have a few minutes. Something has come up and I need to attend a meeting regarding the future of our Logistics Division.

Katie: Does that have to do with your merger with Horizon Electronics?

Steve: Yes, it does. Rumor has it that their transportation costs are 20 percent less than ours and we may be in trouble after the merger. By the way, my people are telling me we should award the West Coast transportation contract to whomever Horizon uses.

Katie: *(Pulls out her PDA and accesses her company's systems.)* Well, Horizon uses RMB Express and although it is true that they are a low-cost supplier, take a look at recent independent industry surveys regarding service quality and timeliness.

Steve: *(Takes a look at Katie's PDA.)* That is interesting. With our emphasis on minimizing inventory, we would be in trouble with those delivery statistics. Where have you guys rated?

Katie: *(Taps her service history button on her PDA and hands it back to Steve.)* We are operating at 99.2-percent on-time shipments for Next Generation, which is the highest in the industry. In addition, we were the only transportation company to earn exemplary status from Next Generation.

Steve: Well, I do admit you guys have done a solid job, but a 20-percent premium is a lot to pay for.

Katie: Why do you think we are at a 20-percent premium? Are you sure that you are calculating a one-to-one relationship between our services and costs and theirs?

Steve: I think I am. What do you mean?

Katie: *(Katie taps the competitive intelligence button on her PDA.)* Based on my knowledge of RMB Express, they are only a transportation company and offer no other services beyond rail capacity. The price they are quoting does not include any value-added services. For example, custom reports and data, warehouse management services, and automatic notifications of delivery problems are at a cost premium over this price. Their best offering is not even as comprehensive as our basic service.

Steve: Well, maybe that's all we really need.

Katie: *(Katie switches to the pricing comparison application on her PDA.)* Let me show you how much the West Coast transportation services would cost if we stripped out the other supply chain management services you requested. I will show you a revised quote from our company and a side-by-side comparison to RMB Express. *(After entering a few pieces of information and requesting a comparison to RMB Express pricing she hands the PDA to Steve.)*

Steve: *(Looking at the handheld)* Wow! You are basically the same price as RMB Express. But what kind of service do I get?

Katie: We will still honor our 98-percent on-time guarantee. You will get the Web interface Next Generation likes so much and the entire line of standard reports we have always given you. In addition, we have a new feature that allows alerts to be sent to the appropriate people in the rare case of delays. These alerts can be sent to a cell phone, wireless handheld, or pager.

Steve: Boy, this could really help in my meeting. Could you print a copy of the RMB Express report as well as this revised pricing for the West Coast bid?

Katie: I'll print it on your office fax machine now. *(Katie taps the Print command her PDA, enters the fax number and a minute later the information begins printing on Steve's fax machine.)* Keep in mind Steve, that if we do not provide the warehouse management services you requested in the original bid, Next Generation will have to staff that function. Horizon has been a target of ours for a while. We have done a lot of analysis of their cost structure from publicly available data. I have printed an analysis of Horizon's inventory costs as well as detailed operational cost so you could see what yours may end up being if you decide to take on all of those activities in house.

Steve: Good point. We don't want to take on those activities. I am pretty confident we can proceed with you on the West Coast business and actually, I would like you to come with me to my meeting to educate my department and Horizon's representatives on what we just discussed.

Katie: No problem. It would be my pleasure.

This example shows the potential impact wireless technology can have on the sales process. The salesperson can improve the ability to cross-sell, deliver greater service, and enhance credibility with the customer. We saw how wireless technology was able to help Katie take a potentially bad situation for her company and change it into a position where she is viewed as a trusted advisor providing solutions for her client. She was able to differentiate her company's offering by having critical information available at her fingertips at the moment of truth. By answering questions and accessing data in the customer's presence whatever is needed to satisfy the customer, the salesperson can close more sales and improve the customer's perception of the quality of the organization he is dealing with. Lastly, with a wireless device, it will be more convenient and less intrusive for the salesperson to update customer and competitor intelligence data than with paper or laptop systems.

The benefits to the organization can be very significant:

Increase sales closure rate. By having the factual information to defend her companies pricing, Katie is able to handle the customer's objections and prove her solution is the most viable.

Increase cross-selling. Using the information on Horizon's operational performance, Katie was able to cross-sell the supply chain management services to Steve as a necessary part of the overall solution.

Improve customer satisfaction. By having the facts in hand Katie was able to help Steve through a difficult situation. Due to the 20-percent lower costs Horizon appeared to be paying, Steve was worried about his position post merger. Katie gave him the information to defend his decisions and prove that Next Generation had an efficient transportation vision.

Enhance breadth and timeliness of customer and competitor intelligence. Meetings like we experienced between Katie and Steve happen everyday in business. They discussed customer (Next Generation information) and competitor issues (RMB Express). Wireless technology has the ability to help sales reps capture this information easily and unobtrusively so the organization can learn through each encounter.

Reduce sales force turnover. Katie has had an experience many sales reps can relate to. Her company helped Katie be successful by having the vision and commitment to invest in tools that aid their sales reps. Katie's loyalty to her company probably increased significantly due to this experience.

Improve sales force productivity. This situation could have taken numerous meetings with many different people had Katie not handled the objections quickly. Rather than spending her time trying to defend her company's quote for the next three weeks, Katie is able to call on new customers and work at winning the Horizon work from RMB Express.

What It Takes to Get There

Leading companies are trying to make the previous example a reality. However, arming your sales force with this type of capability is not a trivial task. We do believe that it is the required and logical follow-on to the investment many companies are making in CRM. In addition, this capability can provide the overall vision for many companies who are spending millions of dollars on one off projects that are designed to acquire specific data and information without understanding how that information can be best integrated into a sales rep's daily job.

Wireless technology is the missing link required to integrate this information into a sales rep's job and maximize the investment in back end systems. The systems that need to be accessed to provide the capability most sales reps would need include

- Pricing and configuration applications
- ERP applications to get at information such as inventory
- CRM systems for contact, prospect, and competitive information
- An Internet connection for general search capability

We've actually designed sales force automation tools that combine these types of information and present that information on a handheld device. The sales reps from the company are able to retrieve customer account information, customer contact information, prior sales history by product sold, marketing programs the company is involved in, a list of assets their customers own (whether their product or a competitor's), and other information. This represents Phase 1 information for the sales people, who previously did not have access to this information. So, these types of support tools can and are being created to facilitate the sales process. In the future, they will only become more feature rich and robust.

As you read the rest of this book and explore the various components of a wireless and handheld architecture, you'll become familiar with the pieces required to make a system like that come to life. In the case of our client, the technologies included handheld devices, wireless middleware to synchronize data, Rapid Application Development (RAD) tools to develop the handheld portion of the system, and back end databases that aggregated all of the necessary information.

Changing the Value Proposition of Products

Every organization is searching for ways to differentiate their products in the marketplace. Adding new features and functions has always been used by businesses to increase sales. It is almost impossible to walk through a store and not notice the phrase *new and improved* on a package. Products ranging from bubble gum to bicycles use added features as a way of differentiating themselves and providing competitive advantage in a very dynamic marketplace.

Wireless technology offers organizations a new method to increase sophistication by adding intelligence into the basic product through customized information and communication capabilities. Wireless technology can provide old products with unique added functionality that actually can change the value proposition of the core, offering opening up new markets and uses that were previously unavailable.

Automotive is one example of an industry aggressively leveraging wireless technology to improve the value proposition of vehicles. The automotive industry has been supplying vehicles to consumers for many decades and, although new models have emerged, the basic function of the car has been to get from point A to point B. Automotive companies have been scrambling to provide consumers with new vehicles that contain the exact options and specifications that the individual consumer wants; this trend is known as mass customization.

Wireless technology enables automotive companies to drive mass customization to new levels. In the future, automotive manufacturers will be able to use a wireless connection to their automobiles to customize the vehicle's content as the consumer's lifestyle or preferences change. *Telematics* is a term in the automotive industry that encompasses this wireless connection to the automobile. New developments in the field of telematics has the potential to position the automobile at the center of an individual's life to act as an information hub that could allow people to manage all aspects of their lives easier. At the heart of telematics is sophisticated wireless technology that is integrated into the vehicle. Within the vehicle, consumers will access telematics systems either through the use of a call button and/or touch screen computer.

In the future, telematics systems will integrate with a personal handheld computer to set preferences and features that consumers want for their driving experience while they are in a convenient location within their home or office. This integration with handheld technology will position telematics as a personal mobility platform that will assist consumers wherever they are. In addition, the integration of in-vehicle computing with the handheld will significantly increase an individual's ability to productively use both computing platforms interchangeably without duplicate data entry or synchronization problems. Let's look at an example on how this technology can transform the vehicle into more than just a car.

A Day in the Life of a Family on Vacation

The Winship family has been looking forward to their trip to Florida for the past year. They have planned a vacation to Disney World to get away from the cold Midwest winters. Each member of the family is eagerly anticipating 10 days of rest and relaxation with Mickey and friends. While the family finishes packing the car, the father, Tim, is inside sitting at the kitchen table finishing the preflight check list. He pulls out his handheld computer and taps the button for the application that interfaces to the in-vehicle telematics system to set his individual driving profile. On his PDA, he selects his destination and other items that he wishes to be available for this trip and locks the door on his way out. Once in the car, his Palm (equipped with Bluetooth technology) synchronizes with the car and updates his driving profile for this trip. They are anxious to get on the road that day because of the long trip and the potential weather problems that can happen almost without warning in March. Tim starts the car and the family is ready to go.

Car: Good morning, Tim. Welcome to the Telematics Customer Response Center. Diagnostics completed and your car is ready to go. How may I assist you?

Tim: Windows down and CD on.

Car: The usual disc?

Tim: No.

Car: Please select artist

Tim: Dave Matthews Band.

(Tim and family listen to the latest new music for the next few hours.)

Car: Construction ahead on Interstate 75. Recommend alternate route using I-475 around the city. Routing assistance or information?

Tim: Routing assistance please.

(Tim follows the onboard computer's recommendation of an alternate route, avoiding the construction. Later, Tim realizes that he forgot to pack the kids' swimsuits and has left his tennis shoes at home. Tim

sets the telematics system to alert him when he is within a few miles of one of his preferred department stores.)

Car: Diagnostic check shows engine is due for a 45,000-mile maintenance check. Would you like me to schedule an appointment at a Ford dealership in Orlando?

Tim: Yes.

Car: Clayton Ford has an appointment for Thursday, which is open on your calendar. Shall I confirm it?

Tim: Yes. Please send an e-mail to my PDA to remind me.

Car: Do you prefer that the dealership pick up the car or to bring it in?

Tim: Pick up.

(After a while of driving Tim presses the Call button.)

Car: Routing assistance or information?

Tim: Information. Local Thai restaurant please.

(Using the global positioning system within the vehicle, Tim's car knows they are 10 miles outside of Knoxville.)

Car: Thai Peppers on Third Street in Knoxville.

Tim: Make reservation and navigate.

(Tim follows the onboard computer's directions and map to the restaurant. The family enjoys their meal and hits the road again.)

Car: Its 7:00 P.M., as you have requested in your profile, a reservation has been made at the nearest preferred hotel, a Marriott in Atlanta. Please confirm.

Tim: Confirmed. Navigate.

(After a restful sleep, the Winships awake and are on their way.)

Car: You have received an e-mail.

Tim: Read.

Car: Message from Grandma, "Tim and family, have fun on your vacation. If you have time on the way back, stop by. We're on the way. Love, Mom and Dad."

Car: Nordstrom's department store is five miles ahead. They have swimsuit and tennis shoe brands you like. Would you like directions?

Tim: Navigate.

(Tim follows directions on the onboard computer.)

(After shopping for their swimsuits and tennis shoes, the Winships continue their drive.)

Car: Air pressure low in rear passenger-side tire.

(Tim pulls over to check the tire and sees the tire is almost flat.)

Car: Weather alert. Thunderstorms with heavy rain are in the area. Caution advised.

(Tim presses the Call button.)

Operator: How may I be of assistance?

Tim: Please send Roadside Assistance for a flat tire.

Car: Roadside Assistance confirmed.

(After the flat is fixed, Tim and family continue their trip. The rain begins to fall and intensify. Tim slows his vehicle down to deal with the decrease in visibility. Unfortunately, some other driver, who decided not to slow down, ends up rear-ending the Winship's van. The other driver's car flips into the ditch while the Winship's van spins out. The Telematics system, monitoring the van detects the sudden change in course and speed.)

Operator: Mr. Winship, are you hurt—please respond.

Tim: Yes, we are fine but the other car has flipped over and is in the ditch.

Operator: Mr. Winship stay calm. We'll send someone immediately.

(The operator calls 911 and directs police and an ambulance to the Winship's car because they know the location of the van. The operator is able to read the sensors from the van to notify Emergency Services that the Winship's have four occupants, the collision happened at 45 mph, and there is damage to the rear of the car. There are no fires; airbags did not deploy; seatbelts were in use; the car is upright; and

no passenger ejections occurred. Thankfully the people in both cars were okay. The next day the Winship family resumes their trip to Florida with a dented rear bumper. Tim presses the Call button.)

Operator: Mr. Winship, how may I help you?

Tim: Well the accident left our vehicle in need of repair. We have an appointment at Clayton Ford in Orlando for our 45,000-mile check, could you send them the information on our accident and make sure they can fix all of our damage as well?

Operator: Certainly.

(The Winships finally make it to Florida and are very excited to get their long due vacation underway.)

Car: ETA to Orlando is 3:30 P.M.; confirm purchase of Typhoon Lagoon tickets and dinner at Pleasure Island.

Tim: Confirmed.

(The onboard telematics computer screen updates to display all types of information on Disney World, their hotel, maps, and information on the city of Orlando. Mrs. Winship uses her handheld device and synchronizes a great portion of this information so that they have it outside of the vehicle.)

Although the previous example is fictitious (and maybe a bit dramatic), the technology to bring applications like these to life is available today. Some of the more sophisticated services that include voice recognition, driving direction, and recommendations will take time to develop and evolve. As you can see, wireless technology has elevated the car from a simple mode of transportation to have many additional features:

- Driving directions and navigation
- Communications through voice, e-mail, and Internet
- Weather alerts and 911 services
- Mobile commerce in reserving hotel rooms and purchasing amusement park tickets
- Concierge services recommending restaurants and stores
- Active maintenance checks on the status of the car
- Other location-based services

The vehicle is now providing information based on the events and lifestyle of the person and transforming its services to fit the personal need of the individual. With the amount of time consumers and workers spend in cars and trucks, it makes sense. Wireless networks and hand-held devices are some of the enabling technologies.

The benefits this type of solution delivers are numerous:

Significant increase in customer loyalty and satisfaction. Personalized services, such as those mentioned previously, will increase the switching costs for consumers and drive up the loyalty most consumers will feel.

Revenue generation due to added services. Consumers will pay for positive experiences. Personalized services appropriately administered have the ability to provide experiences never before possible and generate new revenue streams.

Faster time to market with better products. Increased information on product performance and usage can be fed into the product development process to make next generation products that fit the needs of consumers better.

Less product downtime. Resident diagnostic equipment can detect problems within the product before it breaks down. Repeated problems within a product line can trigger an engineering change to correct a design flaw, reducing the costs associated with recalls and warranties.

Although we focused our example on the automotive industry, there are other areas where wireless technology can be embedded into traditional products to increase the utility of the product and differentiate it from the competition. For example, envision a medical products company embedding technology into drug administering devices that report back to patients, doctors, insurance companies, and the supplier on usage and inventory of that drug. Also, imagine the supplier of electric meters embedding wireless technology that enables utility companies to make on-site meter reading a relic of the past. In each of these examples, the company supplying the breakthrough product would have a distinct advantage over the competition by providing products with wireless tie-ins.

The Bridge from Here to There

Fortunes are won and lost in moments of transition. I think this statement applies to the state of development we find ourselves in with wireless technology today. Firms can embrace this technology and apply it in unique ways to

- Help customers become more connected with the business
- Assist employees with daily tasks
- Alter the core value proposition of their market offerings

Those that successfully execute their wireless strategy will find themselves in an enviable position.

Although all aspects of the technology we discussed may not yet be ready for prime time, many of the productivity-enhancing applications for employees are ready and quite capable of helping organizations reduce costs and protect past investments made in back end systems. The benefits will only be realized by attacking all three of the levers for changing an organization: organization structure, processes, and technology.

The examples presented previously are exciting to think about. Don't let the excitement get in the way of the strategy and ROI processes you must go through first. For example, the sales force example could be a powerful competitive advantage for a company. However, if your sales force is 20 people, you may not be able to cost justify it. It depends on the functionality you need. Make sure the cool factor of the technology does not get in the way of what is possible and what will actually benefit your business.

After three chapters of business, it's now time to turn our attention to all of the wonderful technologies that make wireless and handheld computing possible. We'll cover everything from the 50,000-foot view to details that are both interesting and relevant in building your company's strategy and solution.

Wireless Architecture

We've now spent significant time exploring how to define a wireless strategy and showing examples of business processes that make sense to take wireless. Now it's time to hop in a plane and view the technology islands, which we discussed in the Introduction, that are involved in creating wireless strategies and solutions. In this case, the technology islands are referred to as the wireless architecture. Independent of the particular solution, the 50,000-foot view of the wireless architecture looks dramatically similar. This fundamental understanding of the components of wireless solution is one of the first, if not the first question, clients will ask us. Painting this picture gives IT personnel and business managers alike the ability to put their arms around what the big pieces of a wireless solution are.

In this chapter, we will explore the wireless architecture at a high level and show how data gets from the device itself to the corporate data behind your firewall. In future chapters, we will break out each section in more detail, discuss the challenges and choices involved, and build the knowledge bridges between each of these technology islands. As a previous IT manager for a Fortune 500 company, I sought this type of information to make intelligent business decisions for the company. As a business and technology consultant for the Global Fortune 2000, I'm asked time and again to explain how wireless works so that business managers can wrap their minds around this crazy, exciting, and beneficial market.

The View from 50,000 Feet

Let's dive in and get the architecture picture out of the way so we can discuss it (see Figure 4.1). This picture comes in many flavors by many different hardware and software manufacturers. Wireless telecommunication companies, for example, will typically show more detail around wireless bearer networks. Wireless middleware companies will explode out details related to how their messaging server does its magic. As a point of order: Some companies use the term *middleware*, others use *synchronization server*, and still others a *conflict resolution engine*. Whatever the vendors you encounter call it, the functions it accom-

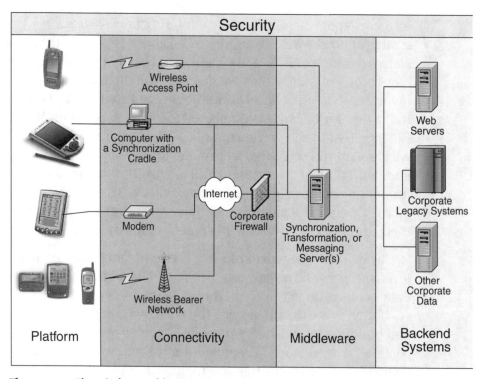

Figure 4.1 The wireless architecture.

plishes are the same. For the purposes of this book, we use wireless middleware, middleware, and synchronization server interchangeably although the word middleware does not necessarily imply wireless technology. There exists many other types of middleware in other computing fields. We'll be covering middleware that works with handheld and wireless devices specifically.

There are five major pieces in the wireless architecture:

1. Platform
2. Connectivity
3. Wireless middleware
4. Back end systems
5. Security

Ninty-five percent of all handheld business solutions have each of these pieces. A minor number forgo security and others have no middleware. Many other solutions may not scale to a size that accommodates an enterprise, but they may have each of these parts. For example, a simple data collection application may involve a single handheld application and synchronization of that data to a single desktop computer. In this case the connectivity, middleware, and back end systems are extremely simple, lightweight, and run self-contained on that PC. Security may not even be a consideration for this solution. Independent of the scale of your wireless solution, all pieces of the wireless architecture must be considered.

Platform

The first piece of the wireless architecture picture is platform. Before we discuss platform, let's define it for our discussion. *Platform* refers to the underlying design of the hardware and software that allows someone to develop applications for it. The PC, however generic the term, typically refers to an Intel processor (or compatible) running one of Microsoft's operating systems. Macintosh from Apple is another platform. They are both hardware and software combinations that developers design and develop against.

In the handheld and wireless space, Palm Computing makes the Palm Operating System. Other companies like Sony, Symbol, and Handspring have licensed this platform. A Sony version of a Palm device will take Sony design cues and have some modified functionality. However, an application that runs on a Palm (the company) device will also run on a Palm-based Sony device. When Palm as a platform is referred to it, it indicates any device running the Palm OS. It may be easier to think of a platform as the operating system of a hardware device. For example, the Windows CE platform, a major competitor to Palm, has been designed to run on devices from Compaq, Casio, and HP.

The Four Platforms

Within the context of this book, there are four major platforms for application or solution: Palm, Windows CE, RIM, and the wireless Web. Although the first three are devices that run a specific operating system, the wireless Web is different. The wireless Web leverages Web technology to send and receive data. Although this does not fall within the definition of platform that I just defined, it will be considered separate in the context of this book. In Chapter 5, "Handheld and Wireless Platforms," there will be a lengthier discussion on the wireless Web and its pros and cons. Because many of our clients talk about Web-based solutions as if they are alternative Palm, RIM, or Windows CE, this book will treat it as such.

Wireless Web technology is typically leveraged in cellular phone handsets. The majority of cellular phones being produced have embedded some type of Web browsing technology, whether that technology is Hyper Text Markup Language (HTML) or Wireless Application Protocol (WAP-based). The combination of a cellular phone and a browser of some sort (or a Palm device and Web browser) make our fourth platform.

It is also worth noting that other platforms beyond these four exist. Some have met with a lot of success and I'm sure they won't be happy being excluded. However from a mainstream enterprise standpoint, these are major platforms we run into with our clients in the States and in Europe. Therefore, they will be the ones mentioned in examples, though most of this information is applicable to all of them.

Device Commonality

Handheld devices possess a selection of common features whether you choose to deploy solutions for Palm, Windows CE, RIM, or some other platform. Cellular phones with their Web-browsing capability do not have many of these features at this time. The current generation is extremely limited in a number of areas in the following, which make them unsuitable for mainstream business applications in many cases. However, this will change quickly as Palm, RIM, and Windows CE are incorporated into cellular handsets. For the other three platforms, common features include:

Good computing power. Although the definition of *good* is certainly up for interpretation, the fact is that handheld devices have the ability to solve general-purpose business problems. Although you won't find three-dimensional modeling abilities integrated just yet, they can recalculate spreadsheet data, view, and edit e-mail attachments, graphs, transmit wireless data, collect data, and run multiple applications simultaneously. Some platforms run 32-bit processors running in the multihundred megahertz range.

Adequate memory. Another term up for debate, especially when viewed from the perspective of a PC. Devices have anywhere from 2 to 64 megabytes of memory with the trend moving to higher amounts of memory. Given the nature of the applications these devices run, that memory is more than adequate to hold thousands of contacts and to do items, voice recordings, multiple applications, product databases, and other corporate information. High-speed wireless network coverage may one day alleviate the need for much of this memory because all information will simply be requested in real time as needed instead of storing it on the device.

Large-ish screens. When compared to paging technology of a year or two ago, handheld devices can display a fair amount of data. The latest wireless devices have high resolution screens that are just a few inches on a side. Some manufacturers are designing handhelds to support different size screens depending on the application. Although you should never (for good reason) compose Word documents on a handheld, you can easily view the sales history for a customer from a Compaq iPaq in full color.

Connectivity options. There are abundant options for getting data on and off handheld devices. Wirelessly is obviously one option. A minority of handhelds have wireless connectivity integrated today. Soon, they all will. Other alternatives include a cradle connected to a desktop or laptop computer and infrared technology for beaming data.

Multiple input methods. On the desktop, a mouse and keyboard are the primary inputs. Attaching a mouse or keyboard to handheld devices, although possible, defeats the majority of benefits you gain by having one. Handheld devices use handwriting recognition, onscreen keyboards you can tap, hardware buttons, and thumb wheels, and voice recognition. In between are a thousand permutations of these basic input options. None are as efficient as a keyboard and mouse, but over time voice recognition may come close. Some of the major differences between devices that run the same operating systems are the input methods that that particular manufacturer has chosen.

Expandability. The original handheld devices of yesteryear had limited ways it could be expanded. Due to the sheer number of vertical markets a device can be used in, industry standards organizations have created expansion options for devices. Most devices manufactured today use a combination of PC Cards, Compact Flash, Secure Digital, Springboard, MemoryStick, and MultiMedia Card technologies to enable for the attachment of GPS hardware, voice recording, wireless connectivity, digital cameras, hard drives, and other items. Figure 4.2 shows a Handspring Visor. It runs the Palm operating system (Palm OS) and comes with an expansion slot for attaching peripherals. On the back is an expansion slot for adding peripherals

Figure 4.2 A Handspring Visor.

called Springboards. Figure 4.3 shows another Palm OS handheld with Magellan's GPS Companion attached to it.

Connectivity

Moving to the second part of picture, you see connectivity. Connectivity, especially the wireless version, is likely the most mysterious piece of the wireless architecture. As we discussed briefly, there are a number of ways for a handheld device to connect to a network or another handheld device whether it is wireless or wired. These multiple paths for connectivity give great flexibility in how information is shared. The purpose of this connectivity is to get data to and from a back end system, which could take the form of a Web site, database, legacy system, or other data store.

Cradles

Cradles are the first and currently the most popular way connectivity can be accomplished. Most consumers and corporate workers attach a

Figure 4.3 Magellan's GPS Companion for Palm V series of handhelds.
Photo Courtesy of Magellan Corporation

plastic cradle to their PC and drop the handheld in the cradle to connect to data. This data connectivity could be to any number of back end systems including

- Corporate e-mail systems
- Internet Web sites
- Intranet Web sites
- Relational databases
- Legacy systems

The desktop or laptop computer, typically attached to a corporate network and the Internet, acts as the gateway between the handheld device and the information it is seeking. Frequently, software on that computer is responsible for determining what data goes to the handheld and what data goes to the back end system.

There is another type of cradle used in enterprise settings, particularly manufacturing and inventory: the network cradle. It appears similar to a regular one and in many ways behaves the same. The main difference is that it connects directly to the corporate network without the need for an intermediate PC because of embedded networking hardware. Figure 4.4 shows the basic architecture of one. In manufacturing and inventory settings, you obviate the need for PCs, their maintenance, and the costs associated with them.

One of the few negatives of a network cradle or cradles in general is their reliance on certain device form factors. For example, a Palm Vx handheld could not fit a network cradle built for a Palm IIIe. The cases of the devices are different. If you throw in Windows CE devices, you have to support even more cradle types. Network cradles will succeed in scenarios where the enterprise of division within can standardize on a particular device type and model.

Infrared

Infrared technology is another connectivity method that enjoys a robust implementation. It has worked well for years because of standards organizations agreeing on how the technology should share data. It's also the way our home remote control talks to the television or stereo system. However, it never really came into its own for transfer-

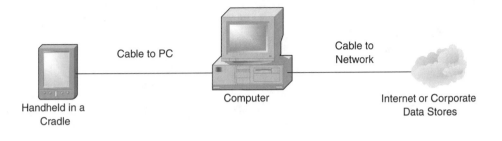

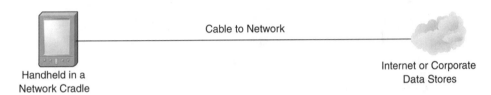

Figure 4.4   Network cradle for communication.

ring computer data. Some laptop and printer vendors embedded IR ports into their hardware, but rarely do people use them.

Infrared enables for the beaming of data between handheld devices, between a handheld device and a PC, and to infrared ports connected directly to a network (see Figure 4.5). In fact, infrared is wireless communication in the truest sense of the word. Unfortunately, in most cases it requires a line of sight to operate and runs at relatively low speeds when compared with a wireless corporate LAN. So, it will likely never become the primary form of wireless communication for any company.

Infrared has enjoyed moderate success in the healthcare space where doctors have the need for updated data, but cannot afford or do not want to afford a wireless LAN. A doctor walks up to an infrared port hooked directly to the corporate network and aims her handheld at the port to synchronize data. Although a network cradle, mentioned previously, would work, it has the limitation of which types of devices you can place in it due to shape of the device and the cradle it must fit into. Infrared works with most any platform and any particular device model.

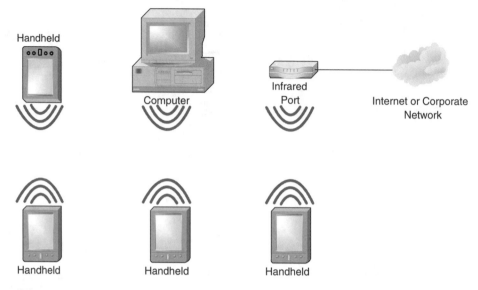

Figure 4.5 Infrared for communication.

Wireless RF

The last method for connectivity we'll explore is true wireless communication, which means using radio frequencies or RF. This is the wireless you think of when you use a cellular phone, pager, or internal wireless corporate LAN. Unlike infrared, this type of wireless does not require line-of-sight for correct operation. Wireless communication can be broken down into three distinct categories: Wireless Local Area Networks (WLANs), Wireless Wide Area Networks (WWANs), and piconets.

WLAN refers to technologies that work over short distances typically within a building or on a corporate campus. The antenna the handheld device communicates with is typically within 300 feet at all times. WLAN is the technology most enterprises are installing in their buildings for wireless network access from desktop and laptop computers. The most common implementation of a WLAN technology is Ethernet or 802.11b and 802.11a. If these terms are unfamiliar to you, we'll cover them in Chapter 6.

WWAN technologies cover networks that are available outside of buildings. They work outside as you roam the city or country and include cellular, paging, and other dedicated data networks. Speeds are typically tolerable or abysmal depending on your perspective. These networks are covered in Chapter 6, as well. To give you perspective, it is these wide area networks that telecommunication companies all over the world have spent billions on to purchase frequency spectrum to build out the high-speed wireless networks of the future.

Finally, we have piconets. Piconets transfer data wirelessly within a short range of other devices. In the specific case of Bluetooth technology, a piconet network can be set up ad hoc between a device and all others within 30 feet. Although Bluetooth is similar to the WLAN technologies mentioned previously, it has some specific advantages over WLAN technology for handheld devices. These will be discussed in Chapter 6.

The vast majority of devices manufactured today, with the exception of cellular phones, do not have integrated wireless technology. Those that do have either WLAN, WWAN, or Bluetooth technology, but not all three. Because of this, a wireless device used inside a building is useless while roaming the country. Conversely, most wireless devices for use while on the go are useless inside a building due to interference. Soon, some combination of these technologies will be integrated into one device, but don't plan for widespread availability of this anytime soon (see Table 4.1).

Table 4.1 A Comparison of Connectivity Options

TYPE	COST	SPEED	PROS	CONS
Cradle	Low	Low–medium	Cradles are typically included with the device.	You must have access to that cradle.
Network Cradle	Low–medium	Medium	No need for a PC.	Certain devices only work with certain cradles.
Infrared	Low	Medium	Infrared is wireless and included with most handheld devices.	Line of sight between infrared ports is needed.
WWAN	Medium–high	Very slow	Wireless connectivity wherever you are (for the most part).	Slow speed; monthly costs for network access.
WLAN	Medium–high	High	High-speed access wirelessly near access points.	No wireless connectivity if you wander too far away.

Wireless Middleware

The other piece of magic is wireless middleware. Middleware, generically, is a software and hardware combination that enables communication. Middleware, like a middleman, sits between two data sources and provides services. Wireless middleware is middleware that provides services specific to the world of wireless and handheld computing. Specifically, services offered by wireless middleware include

- Pushing and pulling data to and from the handheld device.
- Responding to requests or messages from handheld devices or smart phones for data.
- Handling conflicts in data when the data changes in multiple places like the handheld and back end system. We call this synchronization.
- Managing the devices by collecting a device software inventory, providing backup/restore services, and performing electronic software distribution.
- Providing some security services between the device and middleware like authentication and encryption.
- Support for multiple platforms including Palm, Windows CE, RIM, and Wireless Web.

Good packages will provide all of these services. For enterprises, wireless middleware usually resides on a server behind the firewall or with a networking company that has outsourcing services.

Pushing and pulling of data is straightforward and you're likely to be familiar with the terms, but synchronization may be a new term for you. Therefore, we shall define it. At its most basic level, synchronization is a process. It guarantees that two or more data sources contain the same information after completion of the process. Synchronization implements business rules and logic that are decided by the company implementing the technology.

For example, what should occur if the price of a particular product is altered on a handheld by a salesperson in the field and at the same time by a sales manager on his or her PC at the corporate headquarters? This is where the business logic is implemented through the middleware. During synchronization, should the salesperson's price override the manager's price? Should the manager's alteration update the salesperson's price or should something else happen? All of these decisions are part of the synchronization process. Middleware, synchronization logic, and real time messaging are all covered in Chapter 7 in detail.

For now it suffices to say that middleware services come in one of three basic flavors:

- **Synchronization.** This service is provided for users that are occasionally connected. Connected has many definitions. It could mean wirelessly or through a cradle. *Occasionally* refers to the fact that the user does not maintain a wireless connection at all times. It is likely they would be out of wireless coverage at different times for different reasons. Interference from being inside a building is one. Lack of coverage in a remote location is another.

  For these reasons, data is typically *stored* on the device as it is collected or created by the user. At a time more appropriate for synchronization, (such as when the user enters wireless coverage or drops his handheld device in a cradle) the data is *forwarded* on to the appropriate corporate data store. The majority of solutions today run on the store-and-forward method due to not only cost but also the lack of real business need for a wireless coverage.

- **Real time messaging.** If the user is fortunate to have wireless connectivity at most times of the day, he or she may be a candidate for real time information. Real time messaging uses a wireless network in the same way a computer would use a corporate network. For example, once the information for an order is collected on the device, it is submitted immediately to the corporate order system. If the user of the handheld, for example, needs to look up a delivery date for a customer, the request is submitted immediately guaranteeing the most up-to-date information. Contrast this briefly with the regular the synchronization method in which all delivery information would have been loaded on the device once in the morning. The most up-to-date information would not have been available. The question becomes: What is the business need for this data in real time?

- **Transformation.** The last type of middleware service is a transformation. A number of technology companies make software that transforms existing Web pages into pages that can be easily viewed on a handheld device. These packages are typically rules-based and include commands like *Strip out all animations, Reformat large* *headings to a specific font, Ignore frames*, and *Add word wrapping to lines with more than 40 characters*. For reasons we'll explore later in the book, this type of middleware is useful in the short-term. Soon it will go the way of the Dodo bird.

Additionally, wireless middleware typically supports other services in addition to the three types mentioned previously. A good middleware package handles

- **User authentication.** Supports usernames, passwords, and groups for access to the data.

- **Device management.** This feature includes taking inventory of devices, electronically distributing software, and letting managers generate reports.

- **Encryption.** Most of the wireless middleware packages out today support the encryption of data between the handheld device and the middleware software independent of the transport method (cradle, IR, WWAN, and so on). Encryption is crucial to prevent eavesdropping on the data going back and forth.

- **Protocol conversion.** Some pieces of middleware will translate messages and data to work on different types of networks running different protocols.

Back-End Systems

Handheld and wireless computing extends the reach of corporate data and corporate transaction engines. Whether the data is stored on a Web site, mainframe, UNIX server, Oracle database, or any other type of system, the middleware tool you select should be able to connect that type of system. Although the PC applications that use the same databases use all of the data, handheld solution designer will likely only transfer or synchronize subsets of that data. Specifically, only the data that carries time and location sensitivity should be considered in many cases.

Security

The final aspect of the wireless architecture to consider is security. Security, as mentioned previously in the book, is not technology or a set of technologies. Security is a system in which technology plays a role. In wireless, security technology plays a role in each part of the architecture. We'll discuss its role in upcoming chapters.

At this time, security technologies are in their infancy with respect to integration. Few vendors do much outside of authentication to networks and encryption through the air. Securing your applications requires planning and some additional costs to make the solution robust. For example, if you are in the healthcare space, Health Insurance Portability and Accountability Act (HIPAA) regulations dictate many of the security standards your computer healthcare systems must adhere to. HIPPA governs handheld devices in addition to standard PCs and servers.

The Bridge from Here to There

Although it seems a great number of interesting explanations and details have been deferred to other chapters, the wireless architecture's purpose was to fly at the 50,000-foot view. Almost all wireless solutions require consideration to the platform or platforms used, the connectivity method to enterprise data and transactions systems, wireless middleware to implement business logic and synchronization, and the back end system itself, which needs changes in many cases to support handheld devices.

With this basic understanding you should be able to put into perspective any piece of technology a vendor will try to sell to you. Over the next six chapters we will dive to a lower level and understand each aspect of the wireless architecture including design challenges and options in front of you and your organization.

CHAPTER 5

Handheld and Wireless Platforms

S hort Quiz: Who invented the computer mouse?
Douglas Engelbart.
What year did he receive a patent for his creation?
1970.

Consumers and those in the computer field alike typically associate the mouse with the Macintosh computer from Apple Computer, Inc. During 1984's Super Bowl, Apple Computer, Inc. unveiled the Macintosh in one of the best commercials ever made. It was then that the mouse and the revolution of computing with a graphically rich user interface (read: windows) instead of text became mainstream.

In the same way, handheld and wireless computing devices have been rising in popularity since the introduction of the Pilot 1000 and Pilot 5000 by U.S. Robotics in 1996 (see Figure 5.1). Recently, IDC reported that *smart handheld devices* sold 12.9 million units in 2000. They expect that in 2004 the worldwide market for such devices will hit 63.4 million (see Table 5.1).

Before the original Pilots, handheld devices like them had been around for a number of years. Their uses were mostly vertically based and their

Table 5.1 Global Smart Handheld Device Sales

YEAR	SALES/EXPECTED SALES
2000	12.9 million units
2004	63.4 million units

Source: IDC Communication

hardware and software proprietary in nature. Package delivery (think UPS) and inventory are two places where they gained acceptance. However, there were two fairly popular consumer precursors to Palm that sold in excess of 1 million units each: Zaurus by Sharp (popular in Japan) and the Psion series (popular in Europe). It's also worth noting that both devices still exist, but they now live in the shadows of the bigger players we'll discuss in this chapter.

Also in the mid-90s, Apple Computer, Inc. made an attempt to popularize Personal Digital Assistants (PDAs) with the introduction of the Apple Newton (see Figure 5.2). The Newton, even though it ultimately failed, was a technological marvel. It introduced advanced handwriting recognition, grayscale screens, sound, a touch screen, and a robust graphical user interface to the mass, consumer market. It seems that Apple Computer, Inc. was somewhat ahead of its time. In any case, many of the hardware and software design ideas that the Zaurus, Newton, and Psion pioneered are incorporated into today's handheld devices.

The Rise of Handheld Computing

Handheld computing has come into its own over the past two or three years. Devices once considered gizmos or frivolous have entered the enterprise mainstream. Everyone from the standard office worker to traveling sales people and shop floor workers are receiving wireless devices to improve their jobs. A number of factors have contributed to this rise in popularity. Specifically, businesses are now looking to handheld computing to open new revenue opportunities and drive down costs. The main reasons for their acceptance in the corporate workplace today include:

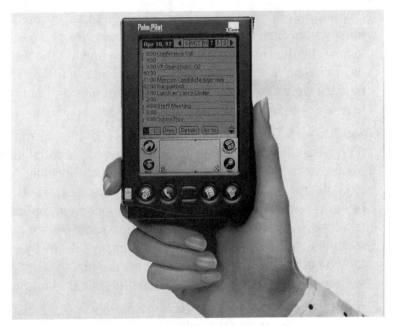

Figure 5.1 The Pilot 1000/5000.

Figure 5.2 The Apple Newton.

Small form factor. The size or form factor of handheld devices has shrunk dramatically. Devices with 200 MHz plus processors with color screens and audio recording exist in packages weighing only a few ounces. For those workers who don't use laptops, the device must be unobtrusive. As a previous Apple Newton owner, the simple fact of the matter was that I didn't like carrying the device around all day because of its weight: one pound. Today's full function color devices weigh between five and eight ounces and are easily forgotten in a shirt pocket, pants, or purse.

Good battery life. In order to replace a tool like a Franklin Planner with a handheld device for the purposes of personal information management, you must make battery life a nonissue. A worker expects instant access to his or her information whether it's on paper or held electronically. The handheld device must operate in the same fashion. Access to relevant information must be fast and available at all times. Palm redefined battery life expectations by choosing not to include energy guzzling technologies like high quality audio and color to focus on the usability issues consumers cared about.

Excellent screen technology. Screen technology has improved markedly over the past few years. Handheld devices continue to receive higher resolution screens while retaining their size. Battery life concerns drop with new energy-saving technologies. Readability in both dark rooms and direct sunlight is improving. Today's devices, because of high-resolution, can display a good quantity of information on a single screen.

Synchronization ability. Connectivity to corporate data is *the* reason handheld devices are adopted into enterprise. Tools that do so, whether connected wirelessly, in a cradle, or through infrared beaming, are a dime a dozen in the industry due to the criticalness of this functionality. Synchronization enables mobile workers and customers to have data they need when and where they need it.

Low price. Dramatic shifts in device pricing make them accessible to a wider audience. Proprietary handheld and wireless solutions for various vertical markets used to command top dollar. Devices could run as high as $3,000 or $4,000. Today a handheld can be had for $200–$400 and industrial strength models go for $1,000–$2,000 with wireless connectivity built in.

Numerous standards. Standards drive cost reductions. Standards create peace of mind for IT directors and CIOs. Handheld and wireless technologies have now past the Wild Wild West portion of their life cycle. Standard operating systems like Palm, Windows CE, and

Research In Motion (RIM) enable development shops to staff people with these skills. Messaging and synchronization standards, such as the eXtensible Markup Language (XML) and SyncML, simplify the interfacing of these devices to corporate data. The industry faces additional evolutions, but it is at a point today where infrastructure investments in most cases will not be a throwaway investment.

Expandability. Almost all of the devices in the industry offer some form of expandability. Although desktop computers are usually upgraded with more memory or hard drive space, handhelds offer new ways to interact with the environment. Accessories such as Global Positioning Systems (GPS), digital cameras, barcode scanners, magnetic stripe readers, and cellular phones open a new world of applications to business.

The Platforms

There are four major platforms worth considering for handheld and wireless computing. It is likely that you've come across them over the past two years. In fact, you may have thousands of these deployed across your organization. The first is Palm, which still holds a commanding market lead in 2001, if you consider all of the licensees of the Palm operating system. Windows CE or the current incarnation's name, Pocket PC, is the second platform. Microsoft has evolved this platform through major redesigns over the past few years trying to find the proper balance between power and ease of use. Now it has a winner. Its market share in the United States has grown remarkably in the past 12 months. Globally, Palm and Pocket PC are becoming the dominant players as well, eroding the market share from other players (see Figure 5.3). Third is RIM; they're the latest players in the market to gain real traction in the United States. Although Palm and Windows CE have tried to be everything to everyone, RIM's objective has primarily been the corporate email market. Because this book's focus is on the enterprise and RIM has gained acceptance there, we'll include it in our discussions as well.

The fourth platform, and it's not truly a platform by the previous definition, is Web-based technology. Some people refer to Web-based applications on handheld devices as Wireless Internet applications or Wireless Application Protocol (WAP) applications. In this book, Web-based

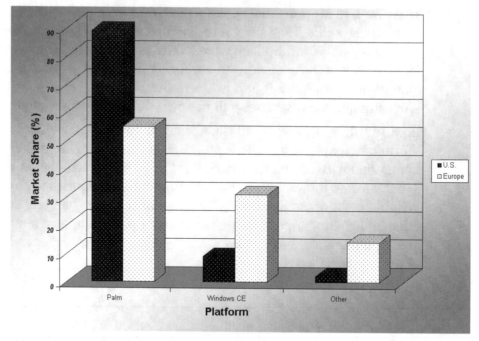

Figure 5.3 Handheld market share.
Source: PC Data

applications will be treated as a separate platform due to the fact that a large contingent of people and technology companies believe that wireless solutions are as simple as installing a Web browser on a device and serving up Web or WAP pages. If this were true, there would be no need for this book. More importantly, there are so many differences between the two types of development styles (Web-based versus non-Web-based) that a large portion of one chapter of this book has been dedicated to its discussion. So, as far as this text is concerned, the four major platforms for development are Palm, Windows CE, RIM, and Wireless Internet.

Other Handheld Platforms

There are two other players with good visibility that I'm excluding from this book: Symbian and Zaurus, both mentioned at the beginning of the chapter. Zaurus remains popular in Japan only. Even there, Zaurus is finding its market share eroding from players like Palm and Windows CE. Similarly, Symbian, popular in Europe and with some of the major cell phone manufacturers, does not have the enterprise presence as a platform as the other four at this time. Symbian has signed some major handset manufacturers as licensees. At this time it's difficult to determine how popular this platform will be globally with business.

Attributes of Handheld Devices

Handheld devices, also referred to as PDAs and wireless devices (even if they are not truly wireless), regardless of the manufacturer have many features in common. What differentiates a handheld from a pager, a cellular phone, and a $100 organizer you can buy from Radio Shack? As with everything else in our gray world, this line is blurring. However, there is a set of attributes that most of the major platforms possess. It is not necessary for them to have all of these, nor is there a worldwide handheld organization blessing one device as a handheld and another as a pager. Let's look at some of these common features:

Operating system. All handheld devices run some type of operating system. In the same way a PC runs Windows and a Macintosh runs the Mac OS to manage the resources of a system, handheld devices too must have an operating system. Palm runs the Palm OS, Pocket-PCs and Handheld PCs use Windows CE, and RIM runs the RIM operating system. If you compare this to a pager or a cellular phone, the software that manages these devices is extremely simplistic.

Large graphical display. Handhelds almost always have large displays capable of showing many rows of text and graphics whether they are black and white or color. If you compare a handheld to a 19-inch monitor, you'll be disappointed, but compare it to a cell phone or pager, and you'll be happy.

Touch screens. Handheld devices beg for large amounts of interaction. To give the maximum flexibility in input, most devices have touch screens on which users input information with a stylus. This is one area in which there are exceptions.

Connectivity. Handheld devices need information to take on the go. That information must come from somewhere. So, most devices ship with cradles. Through this cradle, a handheld can upload, download, and synchronize data with a desktop PC or the Internet. A minority of devices today possesses an inherent wireless capability. Over the next few years wireless capabilities will be built into all devices.

Memory. Cellular phones and pagers used to come with little memory, typically enough to hold a few pages or a couple of caller ID numbers. Handhelds began their lives with little memory, but most today ship with anywhere from 4–64MB of memory making them more than capable to run sophisticated applications.

Programmability. Handheld devices are programmable. Software developers can create general purpose computing applications to extend the out-of-box functionality of the device. They utilize mainstream languages like C, C++, Java, and Visual Basic.

PIM functionality. Handhelds come standard in most cases with Personal Information Managers to track Address, Calendar, and ToDo information as well as transfer email.

Now that we have explored some of the attributes that make up a handheld device, it is time to look at the four most dominant players in the market today.

Palm

The Palm platform currently owns the lion's share of the handheld market, though this is changing (see Figure 5.4). In 1996, when it was released, Jeff Hawkins, inventor of the device, took a road less traveled. Instead of attempting to cram more technology features into a small device so as to mimic a laptop's computing abilities, Jeff and his company took a much different approach. They understood that inexpensive devices with long battery life, data synchronization, and instant access to a person's most critical information would be the successful criteria. They forewent color screens, copious amounts of memory, robust sound, and fast processors. The result created an industry, the handheld and wireless industry.

Since its invention, Palm has expanded its devices to add color capabilities, built-in wireless, and other features. They have licensed their platform to a number of companies including Handspring, Symbol, Sony, Handera, Nokia, IBM, Kyocera, and Samsung. Each company has integrated its own version of value-added features (see Table 5.2).

Figure 5.4 The Palm m500.

Palm Pros

- Palm is relatively inexpensive.
- The user interface is simple and easy to use.
- Palm is the worldwide leader in market share.
- Palm has a wide range of flavors (Palm, Handspring, IBM, Sony, and so on).
- Long battery life.
- Great connectivity options to corporate data.

Table 5.2 Palm Licensee Overview

MANUFACTURER	VALUE-ADDED FEATURES
Handera	Sells a standard Palm-based device with added expansion slots and higher resolution screens aimed at enterprise.
Handspring	Created a Palm-based device with a well-accepted, non-industry standard expansion slot. A number of devices from GPS and cellular phones to digital cameras and barcode scanners may be attached.
IBM	Sells a standard Palm device with a black case directly into enterprise accounts.
Kyocera	Sells a tri-mode digital cell phone with an integrated Palm device.
Nokia	Plans to release a Palm/cellular phone combination. No device has emerged to date.
Samsung	Plans to sell a digital cell phone with an integrated Palm Operating System (OS) and a color screen.
Sony	Created a series of Palm-based devices that use Sony's Memory Sticks and has added multimedia capabilities.
Symbol	Manufacturer of a Palm-based series of devices with integrated barcode scanning, industrial strength designs, and wireless capability.

Palm Cons

- Compared to Windows CE, the devices are under-powered for audio, video, and voice recognition. This should change in 2002 with the release of Palm devices running on Intel's StrongARM processor.
- Most devices lack color for cost and battery life purposes.

Windows CE

Microsoft, never one to concede any fight, has launched multiple generations of handheld devices over the past four or five years. Just as their market share slipped to under 1 percent, they fought back with their current generation of PDA and created a substantial market beachhead (see Figure 5.5). Microsoft's Window CE (also called Pocket PC in its current incarnation) platform is gaining significant market share both in the United States and elsewhere around the globe. Even though its

Figure 5.5 The Compaq IPaq.

global market share is in the single digits or low double digits, it is one of the fastest growing platforms. Unlike Palm, Microsoft does not manufacture devices. They license it to other companies like Compaq, Hewlett Packard, and Casio. Microsoft is also pushing hard to get Windows CE embedded in other devices. They seek to play a role in automobiles, home appliances, and cable set top boxes, to name a few.

The major manufacturers of Windows CE handhelds include Compaq, Hewlett Packard, Symbol, and Casio. With the exception of Symbol, the other devices are similar in many ways. They share slightly different screen technology, memory amounts, and processor types, but they are not as varied in design as Palm devices currently. Like Palm, you can find wireless modems, GPS, and other peripherals for Windows CE. Symbol has also licensed the Win CE operating system. Symbol, again, has added barcode scanning hardware, a rugged case, and optional wireless connectivity to satisfy the needs of a variety of vertical markets.

Windows CE Pros

- High powered devices
- Multitasking abilities
- Integrates with Microsoft's products fairly well
- Color on most devices
- Better suited for multimedia applications

Windows CE Cons

- The Windows CE interface can be daunting for people looking for a personal organizer, and not a small desktop computer.
- Battery life is relatively short.
- Devices are more expensive.

RIM

In any technology arena, this question inevitably comes up: What is the killer application for (insert technology name here)? My belief is that killer applications evolve over time; they are not something you can just think up in most cases. However, it is true that wireless email is an extremely useful application. RIM believes they have perfected it (see Figure 5.6). RIM's Wireless Handheld device design departs from the traditional handheld device. They've included a tiny keyboard and a jog-wheel so that the device excels at sending and receiving email. With each device purchased, a user also pays for email access, Web access, or other data services. RIM operates their Blackberry email service that integrates the device into a Lotus Notes or Microsoft Exchange environment easily. To date, RIM has been extremely successful at penetrating enterprise markets for email. Although their devices can run general-purpose data applications, it has yet to be seen if they will succeed in this area. RIM also licenses its devices and email infrastructure to other companies that would like to host the email service for other customers and brand the device.

Figure 5.6 A RIM 957.

RIM Pros

- Outstanding wireless email platform
- Great connectivity to Microsoft Exchange and Lotus Notes

RIM Cons

- Pricey
- Low processing power
- Unclear as to how suited RIM is to general-purpose enterprise applications

Wireless Internet

Finally, we've come to the Wireless Internet as our last platform. Again, it's probably not a platform in the truest sense, but it is an architecture against which a company can design and implement applications. The Wireless Internet comes in two forms traditionally: An HTML-based Web browser or a WAP-based Web browser. Nearly all handheld devices including smart phones (cellular phones capable of sending and receiving data) either come standard with or have available to them a WAP browser. Most handheld devices whether Palm, Windows CE, or RIM have available to them both HTML and WAP browsers (see Figure 5.7).

It is worth briefly discussing WAP and the difference between it and HTML for Web browsing. In the wireless space, WAP is undoubtedly one of the most popular (or infamous) technologies discussed. There will be a more lengthy discussion of WAP in a future chapter. Figure 5.8 shows a smart phone running a WAP browser.

However, it should be noted now that WAP was designed to take into account the inherent limitations of wireless computing, such as lost wireless data (interference, and so on) and small screen sizes. It utilizes a language similar to HTML that is extremely easy to learn. With WAP, enterprises can deliver Web-based applications to small form factor devices. There is a large contingent of industry pundits and technology companies that believe delivering wireless or handheld applications is

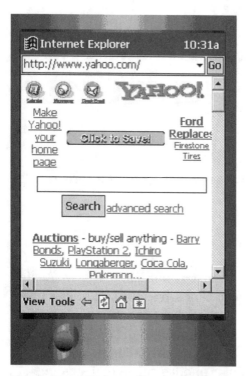

Figure 5.7 A Web page from a Pocket Internet Explorer running on a Windows CE device.

Figure 5.8 A Nokia smart phone displaying a WAP page.

as simple as placing an HTML or WAP browser on a device and serving up pages. However, what they don't discuss are the challenges of operation when there is no wireless coverage or the fact that the browsers do not integrate well with the data and applications already on the device.

Wireless Internet Device Pros

- Leverages standard Web technology
- Relatively easy to set up and roll out to users

Wireless Internet Device Cons

- Smart phones and related devices typically have limited input methods.
- Low processing power.
- Very limited screen displays (on smart phones).
- WAP and HTML are currently ill suited for robust enterprise applications.

Choosing Devices for the Enterprise

In a previous lifetime at a major consumer products manufacturer, I worked in Information Technology reviewing and evaluating new technologies. When you have a large number of employees (100,000) over a wide range of geographies, standardization of technology is necessary to reduce overall IT costs. Long before this enterprise considered business applications, they invested months determining the correct platform to support corporately. Your business will likely choose to go through the same process. If and when you do, you will need to consider some or all of the following factors:

- Out-of-pocket device cost
- Ease to integrate into the enterprise infrastructure
- Tools for remotely managing and updating devices
- Security issues related to the physical device and wireless data
- Ability to source the hardware globally
- Ability to receive a single price globally
- How to handle fulfillment and returns
- Applicability of device to business requirements

Different organizations will always have different requirements. It is my belief that medium and large companies cannot effectively move their businesses forward and continuously innovate if technologies are always distilled down to one choice. It is more effective to find two or three that enable some of the benefits of standardization while permitting the business to choose what is most appropriate for their needs.

This recommendation is even more applicable to handheld devices. Enterprises, in some cases, refer to desktop computers as general-purpose computing devices. No one calls handheld devices general-purpose handheld devices. The small, personal nature of the hardware is typically built to excel at some particular task. RIM, for example, was designed to be a wireless email platform. It performs these tasks very well, which makes its platform popular in a number of spaces. Its lack of a touch sensitive screen makes it ill suited to some other types of business applications. Similarly, Palm does a great job with data capture and synchronization. However, the ability to quickly compose emails or receive alerts is currently lacking in its repertoire. Distilling all of the handheld platforms down to a single type for your enterprise may end up being like fighting the Vietnam Conflict: a long, painful battle you will ultimately end up losing. Handhelds are personal by nature and focused to excel at certain tasks. What will work for your sales force may not enhance other workers' productivity.

Palm or Windows CE

Not to say that RIM isn't a fantastic platform, but this is the $64,000 question I'm asked *all* the time. For $64,000 I'll tell you the answer. Okay, so I'm kidding. I won't ride the fence on this; I'll share my experience. Prior to Windows CE 3.0 and the devices that are out today (also Pocket PC devices), Palm walked all over Microsoft. No enterprise really considered Microsoft's handheld devices because too *many* features were packed in and none stood out, poor performance, mediocre device design, and high prices. Palm, without having color or high quality audio, hit the sweet spot of devices: simple to use, inexpensive, long battery life, and connectivity to almost anything. That is the magic formula that really made handheld computing mainstream.

Many will argue that Windows CE is a Microsoft product, therefore it will integrate better with a company running a Microsoft infrastructure. I'm not sure I agree with this mentality. Palm's devices connect to any data source or transaction system. Whether they can show an Excel spreadsheet as well as a Pocket PC is academic. Who really creates spreadsheets or word processing documents on a handheld? One percent? If you exclude tie-ins like that, both device platforms can talk to any type of back end: SQL Server, Oracle, Notes, Exchange, and so on.

If either device is asked to do general-purpose field data collection, sales force automation, and so on, you'll find that both excel at these tasks. When you get into applications like multimedia (animation, video, and audio) or applications that require serious number crunching, Windows CE pulls ahead for the time being. Both sides have their supporters, and anyone who has used these devices extensively will have opinions on the subject that border on religious.

For ease of use, Palm still wins hands down. Some call the interface elegant, others simplistic. Either way, computer neophytes or those who just don't like handhelds are more likely to use them. I've seen this time and again. Someone will be looking to replace his or her Franklin Planner with an electronic organizer and repeatedly choose a Palm device. Since its humble beginnings, Palm has shown its ability to add hardware and software features to compete with Microsoft. Microsoft has shown its ability (and full bank account) to reinvent itself into something that consumers and enterprises are beginning to adopt in droves.

I don't believe there will be one winner in this category. Microsoft, Palm, and some other platform will each have a fair chunk on this market a couple of years out. Hardware will be parity in features, but run different operating systems. As with other types of platforms, like desktop computing, it will be all about the applications and solutions you can buy or build. Whoever has the better ones will garner the largest user base.

Criteria

I still haven't answered the question of which device to purchase. Again, I don't think there is a particular platform or brand that rises above all the rest. So, we'll look at a variety of criteria. These criteria will help you decide which one (or ones) you should choose. The criteria are many. Additionally, some of the criteria are applicable only under certain conditions (that is, being used with certain vertical markets). For example, battery life may be much more important for a salesperson who is away from the office for a week at a time. Contrast this with a delivery person who will re-charge her handheld device at the end of each day back at a warehouse. Due to the correlation between criteria and business use, these are the most applicable criteria and the general categories by which you can compare devices:

Processing power. Is your business application data captured with a number of digital forms or does it require voice recognition? Collecting data takes little processing power. Voice recognition will take all you can throw at it and more. The processing power of devices varies widely today.

Expandability. Many vertical markets require that some peripheral be attached to the device to accommodate a task: GPS, barcode scanning, digital cameras, cellular phones, and wireless connectivity. Over time many of these features will find their way into devices as default options. Until then, choose devices that support the added technology you need as built-in features rather than add-on hardware. Unfortunately, most devices today only support one add-in hardware peripheral at a time.

Battery life. Battery life can certainly be an issue for your application. True wireless connectivity, whether in a warehouse or while traveling, drains batteries in a hurry. Color screen technology also plays a large role. The vast majority of devices out today have black and white or grayscale screens simply to save battery life. Test the devices you are considering under your real world conditions before committing to quantities of them.

Memory. RAM is also considered a limitation of handheld devices. Most RIM devices ship with 5MB of RAM, Palm with 8MB, and Windows CE with 16MB and above. Excluding multi-media, these devices can handle most enterprise applications you can throw at them. They are all quite capable of searching and sorting through thousands of records.

Also, you cannot do a straight comparison of memory sizes between devices. A 16MB Windows CE device has twice as much RAM as a Palm m505; however, due to the way the operating system uses memory, you may be able to write the same application for both platforms and find they both run equally well. Simply looking at the total RAM is not the whole story when determining which device is better or more capable.

Connectivity options. There are a number of ways to connect devices to corporate data sources, including cradles, infrared, multiple wireless networks, and dial-in modems. Determine what options are available to the device you are considering. If the device can connect wirelessly, how many different networks does it support? You should accept the fact that it will be difficult or impossible to connect geographically separated workers with one wireless network, though the telecom companies would have you believe differently.

User interface. Although you can take user interfaces for granted on any desktop platform, you cannot do so for handheld devices. Some platforms have much more graphically rich interfaces. If you mentally have decided that your data will look the best in an Excel table format on the screen, make sure the devices and development environments support an interface with the ability to grid data. Many times you will choose to design the system one way and be forced to go another route due to some interface limitations.

Middleware support. Almost all platforms will connect with any enterprise data source whether that source is a Web server, XML database, mainframe, or relational database. The question then becomes: How easy is it to do that? Some platforms have tight integration with collaboration systems like Microsoft Exchange and Lotus Notes. Others perform well when exchanging information with Web-based interfaces. Figure out what things you need to connect with the most and choose a platform that can connect quickly.

There are other criteria in addition to these. We'll cover some of the others as we continue through the chapters. As you begin to formulate ideas for handheld and wireless computing, note the non-negotiable criteria for the devices and make sure the manufacturer can meet those needs. Unfortunately, you cannot simply create rules like: when creating a healthcare application, use Pocket PC. Or, if you are creating a sales force automation application, use Palm. It is much more dependent on business processing moving to handheld and wireless devices. The good news is that often times more than one platform will do a suitable job at the same task.

The Future of Platforms

Is Palm going to win? Windows CE? Some other yet to be released platform? Cellular phones? This question is the one I'm asked more than any other. Obviously no one wants to bet on a losing horse. Unfortunately, it is far too early to make that determination. I am willing, however, to give you some ideas of where the market is likely to go.

Cellular phones and handheld devices will converge for a great portion of the population. There will be outliers who continue to use old cellular or handheld technology, but a great many will adopt devices that have both voice and data capabilities. Handheld devices need a wireless connection going forward. Cellular phones don't necessarily need data, but because consumers and enterprises will own them, it only makes sense to have data capability.

Palm and Windows CE will be around for some time. Palm will likely not dominate forever. Windows CE and RIM are eating at its market share, which is to be expected when you own most of the market. I don't think there is a need for just one platform to be around. If applications of the future are designed according to the future wireless architecture I outline in Chapter 9, there will be little need for any one plat-form to dominate.

For business this adds difficulty because many like to standardize on a single platform. Currently the platforms are too immature to pick one, even if you had to. Palm excels at certain things Microsoft can't do and vice versa. Choosing a single platform to fulfill the requirements of sales teams, executives, and warehouse workers is next to impossible.

Finally, handhelds are much more personal to users than laptops. Get-ting everyone to use one platform will always be a challenge. Addition-ally, manufacturers will continue to create devices specific to different niches. RIM has started this with devices that excel at email. Who is to say that Palm, Microsoft, or another player won't release a suite of handheld products completely geared at healthcare or transportation and logistics? These devices may have hardware for factors specific to the market they're going after. A doctor's handheld may require a port to hook up devices to record vital signs or hardware buttons to call up lab results.

The Bridge from Here to There

This chapter considered the four platforms for running wireless appli-cations. We discussed three true platforms (Palm, RIM, and Windows CE) and a fourth, Web-based technology, for delivering business func-tionality. We also examined the criteria by which you choose platforms noting that some of the platforms perform equally well across a range of tasks. If you want device-specific information and differences between platforms, I recommend visiting the Web site of the companies listed previously or buying one of each type of device. Many times nothing substitutes for laying your hands on actual products.

Our next stop is the wireless technology primer. We need to take a look at the various wireless networks available to enterprise including cellu-lar, 802.11b (Wi-Fi), and Bluetooth. The wireless landscape seems foggy and confusing. Chapter 6 will alleviate some of this confusion and sepa-rate fact from fiction related to wireless networks.

CHAPTER 6

Wireless Technology Primer

The Wireless Technology Primer is pretty much a staple with all wireless books in print today. The question I wrestled with was: How detailed and technical should the information shared with information technology managers be? One argument says that you should not care how your data gets from the device to your corporate data stores. All you need to know is whether or not coverage exists for your workers, how much wireless data costs, and what types of application can you make. On the flip side, many people want to understand, if in layman terms, how the underlying technology operates. The main reason is that it makes us capable of understanding how future technologies that are introduced fit into the bigger picture. In the spirit of islands and bridges, this will be a slightly more technical discussion than not.

This chapter is concerned with the basic concepts of wireless networking technology including terms like frequency, protocol, and packet-switched. We then look at the various WWAN technologies including cellular phone standards and data-only networks. Indoors, companies are beginning to use WLAN technologies like Bluetooth and 802.11b. If there is a chapter in this book you can skip and be no worse for the wear, this would be the one. However, this chapter will serve to clear up many of those crazy acronyms you hear each day and create a clear picture in your head of true wireless communication and how it is used.

Concepts

Before we can discuss 2G, 3G, cellular, and other technologies, there are some fundamental terms that we must go over. None of them are particularly complicated and they are terms you will read time and again in other books and magazine articles. I picked these terms out of many because they explain at a basic level how wireless data works and it also dispels any myths that wireless data technology is somehow different than an FM radio broadcast or how a cordless phone works.

Spectrum/Frequency

As a quick reminder, frequency refers to the number of times per second an electromagnetic wave oscillates. The measurement for frequency is Hertz (Hz) (see Figure 6.1). If a wave oscillates 4,000 times per second, it has a frequency of 4,000 Hz. As you also may remember, the human ear typically hears in the frequency range of 20 Hz to 20,000 Hz. It just so happens that your eardrum and brain interpret frequencies in this range as sound. Sounds closer to 20 Hz are deep and low; sounds near 20,000 HZ or 20 kHz are high-pitched. To be clear, sound is a mechanical frequency because it requires air to propagate. In other words, you can't just increase the frequency of sound and expect to get light. They have different physical properties. This example is simply being used to illustrate the differences in frequencies and how someone or something may interpret them.

An area of the electromagnetic spectrum that our body can and does detect is visible light. Frequencies for visible light are in the trillions of hertz. Other types of everyday waves that fall in the electromagnetic spectrum include

- FM radio
- Cellular phones (analog and digital)
- Walkie-talkies

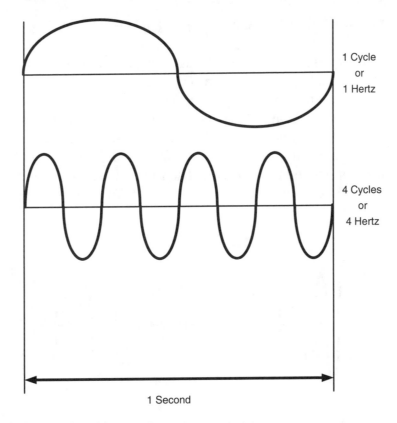

Figure 6.1 Examples of frequencies.

- Cordless phones
- Bluetooth
- HDTV broadcasts
- Police radios
- Television
- X-rays

Governments around the world control this spectrum and dole it out for different uses. The takeaway here is that wireless voice and data communication is no different than any other form of wireless communication and because the frequency spectrum is extremely crowded, carriers are spending huge sums of cash (over $100 billion) to own spectrum for current and future cellular and data networks.

Transmitting Information

Just because you're transmitting a frequency does not mean that you are transmitting information. This signal must be altered over time or *modulated* to send information. Three common ways data is transmitted using electromagnetic frequencies include amplitude modulation (AM, the technology AM radio uses), frequency modulation (FM, the technology FM radio uses), and pulse modulation. Graphically, these are represented in Figure 6.2. The first electromagnetic wave varies in its height only, which we call amplitude. You can correlate the ampli-

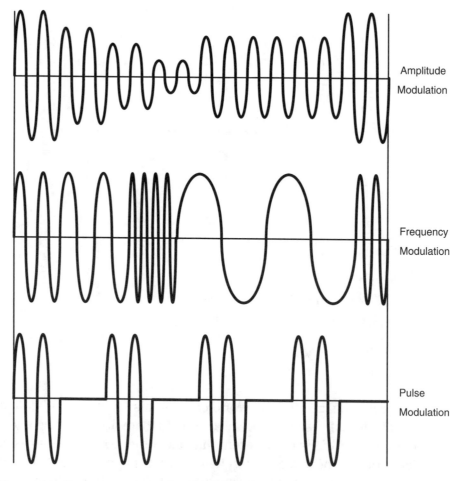

Figure 6.2 Three ways of encoding information wirelessly.

tude with how intense or strong the signal is. Equipment that encodes information whether voice, data, or music using AM varies the amplitude of the frequency over time while holding the frequency constant to transfer data.

In the second part of Figure 6.2, we vary a different aspect of the wave, the frequency. If you notice, the height or amplitude of each frequency remains the same over time. By changing the frequency (Hz) we can also transmit wireless data. FM is the basis for both FM radio and cellular phone technology. Finally, we have pulse modulation, which is seldom used. Pulse modulation is simply encoding information by turning a constant frequency with constant amplitude off and on over time.

That's all there is to it, more or less. Wireless communication of all kinds relies on getting voice, traditional data, video, radio, and so on from one location to another. With the use of amplitude, frequency, and pulse modulation, we have methods for communicating.

Protocols

Once a manufacturer or wireless equipment vendor chooses a method of wireless communication medium, they need to choose a format (protocol) for this data so that the two participating wireless handheld devices can understand what is being transmitted. A protocol is a series of steps that must be followed in a certain order to accomplish a task. This simple definition applies to protocols in real life as well as the world of computing. Take the purchase of coffee from Starbucks as an example. If you want to order coffee, there are a series of steps that must be followed in a semirigid order for you to successfully and happily walk out with Caramel Macchiato (if you've never had one, go immediately to Starbucks and order this). The steps are as follows:

1. You: Walk up the cash register (with a happy Starbucks employee working behind it).

2. Clerk: What would you like to order?

3. You: Caramel Macchiato.

4. Clerk: What size?

5. You: Medium ("Grande" in their lingo).

6. Clerk: That will be $37.50 (give or take).

7. You: Hand cash over for payment.

8. Clerk: Thank you. Go to the pick-up counter to receive it.

9. You: Walk to pick-up counter and receive Caramel Machiatto.

At some level, this ordering process is a protocol. You can't skip steps. Without telling the clerk the size of the drink you want, they can't make it. If you skip paying for it, the police arrest you. Moreover, you can't change the order the steps. Walking in the store and going immediately to the pick-up counter will garner stares from everyone. Handing cash to the clerk before ordering, while making the Starbucks employee happy, will not necessarily get you a refreshing coffee drink.

Protocols for the transmission of wireless data are no different. They dictate the steps involved in transmitting data: You must send the name of recipient, sender, date, and time along with the data itself. It also mandates the order of the steps: Send the recipient name first. Wait for an acknowledgement that the recipient's cellular phone is on. Send the date and time of the message next, and so on. It's also because of protocols that we experience the challenges of having a cellular phone or wireless Palm device that can be used in any country or geography. Later in the chapter, we'll look at the three major cellular technologies along with data-only networks and discuss pieces of their protocols.

Circuit and Packet Switching

Now, we'll look at circuit and packet switching. To this point we have discussed the frequencies information are transferred on, how to encode data within a frequency, and the need for protocols to order and manage that data in the air. Two other terms that need to be defined are circuit switching and packet switching.

Circuit and packet switching are used to describe one of the inherent properties of a network whether it is wired or wireless. It's best described with an analogy to our tried-and-true telephone network (plain old telephone service [POTS]). Before AT&T and other traditional telephone companies went to digital systems for transmitting voice, there used to a be a real, physical, and dedicated connection between your phone and the phone of the person you were calling. You spoke over your own personal copper wire—your own circuit.

No, the phone company didn't string cables from your house to every other house in your town. Instead, all the cables were strung to a huge switchboard in your town. In the very early days, you picked up your phone and told the operator to whom you would like to speak. She then plugged in a wire that connected your line to the line of the other person. For that phone call, the two of you had a dedicated piece of copper wire—a *circuit*.)

Similarly, wireless connections can be circuit-switched. Obviously there is no wire, but there is a process to establish a dedicated connection between your handheld device or cell phone and whatever you're connecting to. Circuit-switched technology has the advantage of a dedicated line and bandwidth because you don't share that with anyone. On the negative side, the time from when you initiate a connection to when you actually transmit data can be many seconds. Also, fewer users can be supported simultaneously because none of the connections are shared. This translates into lost revenue for a carrier. There is a trend in the industry as carriers shift from circuit- to packet-switched technology.

If circuit-switched technology is like your own private road for driving, packet-switching technology is a road that everyone shares. Packet switching enables for a more optimal use of an air channel because everyone's cellular phone or wireless handheld device shares it. Continuing with the analogy, when too many cars enter the road, you get congestion. The same is true of a packet-based network. Too many simultaneous users degrade the quality and performance for all (see Figure 6.3). In this diagram, packets from cellular phones and landline phones are combined at different points to share the bandwidth. At certain points the cellular phones grab packets assigned to them, and routing equipment sends certain packets to landline phones.

The other advantage of packet-switched technology and another reason why network operators are migrating in this direction is that packet networks have an *always on* feature. Creating a connection to the network is easy and fast. It lets users feel like they have instant access instead of waiting for a connection like you do when you dial in to your ISP with a modem from home.

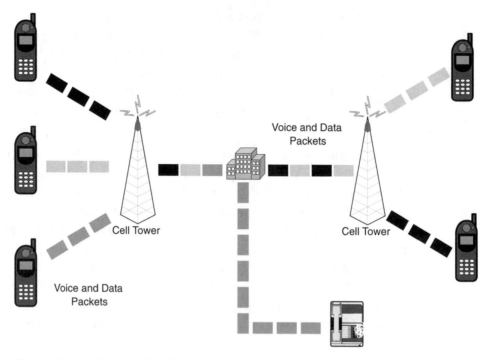

Figure 6.3   Packet switching diagram.

Cellular Communication

I always enjoy when a technology term becomes part of our consumer mainstream. I've heard home computer users ask things like: What's the URL? What POP3 server do you use? How many megahertz did you buy? The term *cellular* is another term that's entered our every day conversations. What does cellular mean and how does it relate to CDMA, TDMA, and other wireless terms we hear day in and day out?

Imagine for a moment that you are planning the wireless coverage of a city for voice and data communication. You have seven frequencies you can use and each frequency supports 10 callers. One option would be to set up powerful antennas in the center of a city and have them blanket the surrounding area with wireless coverage. Under this scenario, you would be able to handle 70 simultaneous voice or data calls in that city.

As an alternative you could choose to break up the city into geographic segments or *cells*. In Figure 6.4, cells are depicted as hexagons. Imagine

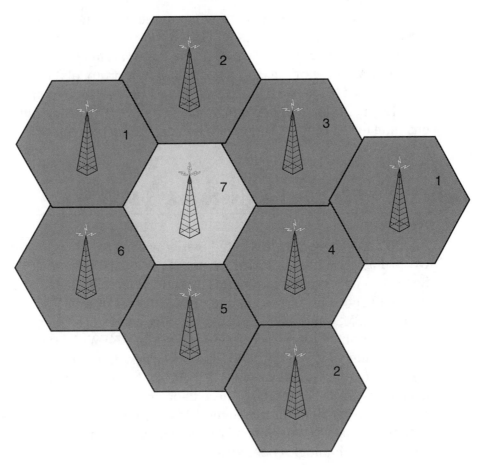

Figure 6.4 A representation of a cellular network.

that each antenna (or base station) at the center of the cell covered a much smaller area, maybe five miles in each direction. The numbers, one to seven, represent the particular frequency a cell site uses to communication with cellular phones. In the first scenario, the antenna covered the entire city with all seven frequencies. In the cellular example, each cell site uses one frequency. Because two antennas using the same frequency are never put adjacent to each other, interference is not an issue. Each cell site can handle 10 simultaneous calls. Let's say it took 25 cell sites to cover the city. Although no particular area could handle more than 10 calls, the overall network supports 250 callers versus the 70 in our original design.

This second example is how our wireless telephone network is constructed. It uses a cellular model. So, *cellular* does not refer to a particular technology. Rather, it covers the layout or architecture of the network. All three of the cellular standards (or protocols) in use today (CDMA, TDMA, and GSM) use a cellular model to handle communication. Each of these standards operates in its own frequency range enabling for all of the standards to coexist within one city or geography.

Data and Voice Wireless Networks (WWANs)

As you are well aware, the United States, as well as other parts of the world, has a multitude of wireless networks available to subscribers. Some of the marketing buzzwords, brands, and technology terms you've likely come across include CDMA, 1G, Mobitex, 1XRTT, TDMA, Verizon, GSM, GPRS, EDGE, 2G, 2.5G, PCS, and Reflex. All of these terms describe Wide Area Wireless Networking (WWAN) technologies. Said another way, WWAN includes all networks that are available while traveling outdoors. Overall, I think some of these companies like to keep enterprises confused by these choices so they can be viewed as the one solving all of the problems. This section is devoted to looking at these terms, what they mean to you, and a bit of information on how particular technologies operate.

Technology Generations

First, let's look at technology *generations* of cellular communication. Depending on where you live in the world, this cellular phone technology may be referred to as 1G, 2G, 2.5G, and if you're lucky enough to live in an Asian or European test market, 3G. *G* stands for Generation. The cellular community refers to each major evolution in cellular technology as a generation:

1G. First generation technology represents the older analog cellular phones from several years ago. This obsolescence is why you don't hear this term used very frequently. The technology is circuit-switched, prone to interference, can handle few calls simultaneously per cell site, and contains no security between the phone and cell site. Most digital cellular phones contain a fall-back analog mode should digital coverage not be available. Sometimes these phones are called dual-mode.

2G. Second generation technology is also considered the first generation of digital technology. It is the technology almost every phone being sold today is utilizing. Three major, digital, cellular 2G standards exist in the world: CDMA, TDMA, and GSM (all described later in the chapter). Connections for this 2G technology are still circuit-switched, but are less prone to interference, require less battery power, have some capability to transfer wireless data, and use some form of encryption to protect the call or data through the air.

2.5G. 2.5G is an extension of existing 2G technology. It adds some features like packet-switched calls and data instead of circuit switched. Recall that packet-switched networks set up connections more quickly and can generally accommodate more simultaneous calls per cell site. Also, data rates for sending and receiving wireless information is greatly enhanced (the interesting aspect to business). 2.5G is an interim step before cellular companies begin rolling out 3G systems. 2.5G technologies are available today in Asia and Europe while the United States continues to wait for theirs.

3G. Third generation is the Holy Grail of wireless data communications or complete marketing hype depending on whom you ask. 3G technology promises high-speed voice and data networks rivaling speeds of computer networks and many times faster than a standard modem connection. Besides cellular phones and handhelds, you would be able to walk around outside with a laptop and a 3G wireless card and still have high-speed Internet access. This is an extremely compelling scenario, if telecommunication companies can even come close to having the technology meet the hype. By the time this book is available, Japan should have a 3G network for general availability. Europe is also beginning trials, and the United States won't see commercial availability of 3G until 2002 or 2003. However, by the time you read this book, some of these technologies may be available to you in a limited trial capacity.

Each of these generations of communication technologies apply to the three major cellular protocols: Code Division Multiple Access (CDMA), Time Division Multiple Access (TDMA), and the Global System for

Mobile Communication (GSM) (a version of TDMA). Each cellular operator has adopted one of these technologies. For example, Verizon Wireless uses CDMA technology for transmitting both voice and data. Verizon is currently using the 2G of CDMA technology, which supports digital communication and some wireless data transmission. CDMA, TDMA, and GSM each have one or more technology upgrade paths to get from 2G technology to 3G.

Also, it is worth mentioning that cellular technology for wireless data isn't the only method of wireless communication. In fact, there are a host of two-way data-only networks available in the United States. The majority of wireless and handheld devices use these data networks because they've been in existence longer and until recently had much better coverage of major cities than cellular technologies. To highlight some of the major data-only networks, look at Table 6.1.

Wireless Network Performance

How fast are these networks? Not fast at all. Because of this, there is a misconception that business or mobile commerce can't be conducted today with wireless technology. This could not be farther from the truth. If your wireless product, service, or enterprise application relies on transferring large data sets with graphics, you could be in jeopardy and discover little adoption by customers or workers. However, if you optimize transactions and keep them focused on requesting and submitting only the data required, there are dozens of enterprise applications or services you can offer wirelessly.

Over the next 12 months, network operators will rush to upgrade their networks from today's 2G technology to 2.5G technology. Many 2.5G technologies switch the network from circuit-switched to packet-switched. However, the more interesting outcome will be support for higher data rates. As I've stated in other parts of the book, 2.5G networks will give wireless and handheld devices all the bandwidth that they need with the exception of a few high-bandwidth video-streaming applications. The real advantages of enterprise data and mobile commerce services will be exploited with these networks, alleviating the need to wait for 3G. Table 6.2 highlights some of today's networks and their speeds.

Table 6.1 A Few of the Data-Only Wireless Networks

TECHNOLOGY	DESCRIPTION	PRODUCTS
Mobitex	Mobitex is the marketing name for Cingular's data network based on Ericsson technology. It is a packet-based network with good to excellent coverage in most medium and large cities.	Palm VII and VIIx, some RIM Blackberry devices.
CDPD	CDPD stands for Cellular Digital Packet Data. Specifically, this is a digital technology addition or upgrade to the original analog cellular networks enabling for wireless data access.	Omnisky modems, many PC Cards for laptops.
DataTac	DataTac is another Motorola technology. It is another one of the original packet-based, data-only, two-way networks. Different versions of it exist for the different continents. Many companies, like Motient, license the technology for use on their network.	RIM Blackberry devices.
ReFlex	ReFlex is a two-way paging network designed by Motorola. A number of paging companies use this technology as their basis for operation.	Various two-way pagers.
Ricochet	A wireless network that relies on network antennas hung from light posts and street signs within select cities. It supports 2.5G-like speeds today.	Various Ricochet modems for laptops and handhelds.

Table 6.2 A List of Speeds for Various Wireless Technologies

TECHNOLOGY	SPEED
CDMA	14.4 kbps
TDMA	9.6 kbps
GSM	14.4 kbps
Mobitex	8 kbps
DataTac	14.4–19.2 kbps
CDPD	19.2 kbps
Re/Flex	9.6 kbps
Ricochet	128 kbps

Network Throughput

Looking at these numbers can be a bit deceiving. Data throughput cannot be determined by simply taking the number presented and dividing by eight to determine kilobytes per second. All data transmissions contain some amount of overhead in addition to the data being sent, things like address information (where the data is headed) and error correction information (for detection of changes during transit). These bits, though necessary, waste bandwidth that could have been used for data. With a standard modem connection, it is not uncommon to have overhead information represent one-fifth of all data transferred. With wireless data, there are many other ways data can get lost or changed. The result: Up to 50 percent of the following numbers may be dedicated to overhead. An already slow connection becomes slower.

Also, when someone tells you that a particular network will let you send data at a certain speed, that number is called the *rated air speed*. These numbers represent the top speeds you *could* see if all the planets align and you are standing still on a sunny, windless day on the top of a hill in a city. That may be a bit dramatic, but in truth, you will almost never see the speeds quoted to you. There are several factors.

First, the physical conditions under which a wireless network operates in the real world are quite different than those in the test lab. For example, there aren't likely to be snowstorms in a test lab and a vicious snowstorm in your area will hamper communication between your device and the antenna. This would also apply to other forms of weather phenomena.

Second, no wireless network operates at its rated speed continuously. Instead, they operate at those speeds in bursts. A perfect example of this is your automobile. Sure, it *can* go 130 mph, but unless you're driving a specially designed racecar, your automobile will have a substantially shorter operating life if you consistently drive at those speeds. The engines of almost all passenger automobiles (at least those sold in the United States) are tuned for maximum efficiency and life at 50 mph. Likewise, all wireless networks have an optimally efficient speed associated with their physical construction that is likely to be less than the maximum.

Finally, allowing a particular individual to attain a high wireless speed means that the user is utilizing more of the frequency spectrum. Wireless carriers are faced with this decision all the time. Do I support more users, each of which could be grumpy because the network is slow, but increase corporate revenue? Or do I support less users, each of which would give glowing reports about our service for performance, but lose out on additional revenue? If the decisions they tend to make about the quality versus the capacity of their voice networks are any indication, you shouldn't expect to operate anywhere near the rated speed without paying a significant surcharge.

Network Latency

When evaluating wireless networks, there is one other factor besides speed to take into account—*latency*. Because of the specific design characteristics of each wireless network, the amount of time a single packet of information takes to make a round trip between the sender and receiver will vary. This term is called *latency*. Two data networks can have similar data speeds but very different latency characteristics. This will make one network look a lot faster or slower than the other will. For example, CDPD has a higher speed than CDMA. However, CDMA has much better latency characteristics. The result is that most applications will appear to run faster on a CDMA network than on a CDPD network. Because many of these networks are available in many of the same areas of the world, you will have to choose between them. A good rule of thumb is this: For most wireless enterprise applications, a speed of 9.6 kbps is sufficient. The determining factors in your evaluation should be latency and coverage.

Coverage and Interference

One of the largest hindrances to companies rolling out true wireless solutions to an entire sales force that is geographically dispersed is coverage, or rather lack of it. Although television commercials from the major carriers will tell you that their networks cover 90 percent plus of the U.S. population, they don't tell you that it represents just a few percent of the entire U.S. land mass. That may not be critical if your employees, customers, and suppliers are in the top 300 or 400 markets. You will find outages outside of all cities except the most heavily trafficked roadways. This is no fault of the carriers. The United States is simply a big place. This is one of the reasons why we discuss the design of hybrid solutions that work in and out of wireless coverage. Also, remember that having cellular voice coverage does not necessarily imply data coverage. On the up side, if one carrier does not have data coverage in a particular area, there is a good chance one of their competitors does.

Another aspect of wireless computing that can be as frustrating as coverage is interference. Interference affects our television programming, radio broadcasts, and any other communication technology (wired or not). WWAN technologies suffer if the user walks too far into the interior of a building. Elevators and basements are also great places to find interference. The funny thing is that you can be in one spot inside of building with no coverage, walk five feet, and have a perfect connection. As a rule, do not count on a wide area wireless provider to have coverage for business processes that occur indoors and rely on wireless technology.

Later in the chapter, WLAN technologies are covered. These are designed for the interiors of buildings and corporate campuses. Although they provide great, high-speed coverage in small areas (compared to the WWAN technologies covering the United States), WLAN technologies also suffer from interference and coverage holes. Typically, this means putting up extra wireless access points. Because of the funny nature of interference and how it's affected by building structures and materials, you cannot simply figure out your square footage and install the appropriate number of access points. Plan on having a site survey by a company that can walk around and measure signal strength at various points.

Emerging Technologies

The world of wireless communication is evolving at a dizzying pace. Japan already has a nationwide 2.5G network in place operating at 64 kbps. Europe has introduced its own 2.5G technologies boosting speeds for all. The United States is poised to begin its rollout. However, how we get from where we are today (2G) to where we're going (3G) is unclear and definitely not a straight line. Technology terms like EDGE, GPRS, 1x, 3x, 3xEV, and others are emerging and it's clouding the landscape not to mention confusing businesses everywhere. Believe it or not, these terms aren't that difficult to grasp and all of them are related to one of the three cellular standards: CDMA, TDMA, and GSM.

TDMA: Facts and Future

Time Division Multiple Access (TDMA), like CDMA is complex and advanced technology for mobile communication. TDMA is a generic method for transporting bits of information. From TDMA come specific implementations of this technology. GSM is based on TDMA, but has many additional services and features that enhance TDMA's basic architecture, including higher quality calls and the ability to transmit wireless data. When people in the United States refer to TDMA as a specific thing like AT&T wireless support TDMA, they are referring to the IS-136 implementation of TDMA. Said another way: TDMA is a generic term; GSM and IS-136 are specific implementations of the technology.

That preciseness aside, let's look at how TDMA does its thing. Most of what you need to know is in the name itself. First, Multiple Access (MA) refers to the fact that more than one user can share a channel. How this is accomplished is through Time Division (TD). TDMA divides a wireless channel into time slots similar to a multitasking operating system works. The CPU of the system makes you believe it is running multiple tasks simultaneously by switching between them at an incredibly high rate. Each process gets a fixed slot of time, and all tasks take turns. One TDMA channel can handle three calls. Each call gets a small, fixed slice of time. This happens so quickly that the caller believes his communication is continuous.

In the case of the IS-136 implementation of TDMA, one channel of spectrum is split into six time slots with two time slots being used for each voice or data connection (see Figure 6.5). For reference, GSM takes a larger chunk of spectrum and splits it up into eight time slots. TDMA (generic) figures out how to break apart voice conversations into these packets, compress them, send it over the air, decompress the packets once they have been received, and reassemble them at the other end. TDMA can handle more calls in fixed amount of spectrum than older 1G analog technologies, which makes it more efficient. Because of the way the air channel is used (time divided), you can calculate the exact number of simultaneous callers that can be handled. This applies to both IS-136 and GSM. As you will see in the next section, CDMA capacity for a given slice of spectrum is not so defined.

As TDMA (specifically, both the IS-136 and GSM implementations) progresses from 2G to 3G technology, there are a few different paths to get

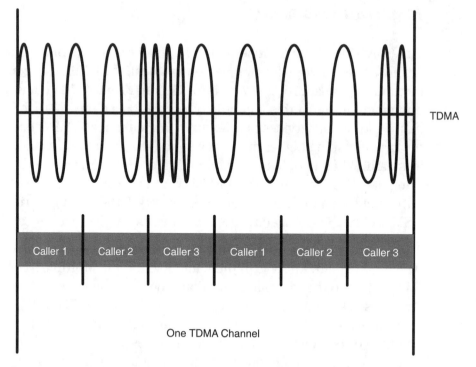

Figure 6.5 A graphical representation of TDMA (IS-136).

there. Each has pros and cons in the technology, cost, and timeline. Many other authors have written articles, white papers, and books on the subject, so I encourage you to read those, if you are interested. See the resource listings at the end of the book for some of these. However, I will highlight a couple of popular buzzwords that you've either already heard or will hear about in the near future:

GPRS. General Packet Radio Service (a 2.5G technology) can increase data speeds of TDMA networks to over 170 kbps. From a practical standpoint, most subscribers will have access to 64 kbps, which is over four times faster than regular GSM.

EDGE. Enhanced Data rates for Global Evolution (a 2.5 technology) can increase data speeds to 384 kbps.

It is unclear if IS-136 and GSM network operators will use GPRS, EDGE, or both to evolve their networks to 3G. Without getting into technical detail, an operator could choose to implement one, both, or neither on their way to 3G.

CDMA Facts and Future

Code Division Multiple Access (CDMA) is in use all over the world similar to TDMA. Also like TDMA, CDMA is a generic method for transferring bits. Multiple Access (MA) has the same meaning. The fundamental difference between the two is how CDMA enables multiple users to have access to the spectrum it uses. CD stands for Code Division. To quote Qualcomm's Web site for an analogy

> "international cocktail party — dozens of people are in the room, all talking at once, and all talking in different languages that you don't understand. Suddenly, from across the room, you hear a voice speaking in your own familiar language, and your brain tunes out all the background gibberish and locks onto to that one person. Your brain understands the "code" being used by the other person, and vice versa."

> Source: www.cdmatech.com

What's being related is that in CDMA, all cell phones are communicating simultaneously. Each voice packet has a *To* and *From* address attached to it representing the phone the voice packet came from and the phone it's going to. Your cell phone only listens to the ones addressed to you, filtering out the rest as noise. This is not too different from the way the computers connected to your office LAN talk to one another. Obviously too many callers on one channel could eventually cause connection issues.

Capacity within CDMA is a whole different story than TDMA. TDMA has a fixed number of users it can handle because a channel is split into time slots. With CDMA, more and more users can be added with the overall quality of each individual connection decreasing. So, it's up to the telcos to decide what quality level they would like to give to their subscribers. One argument is to take all the revenue you can and understand that you'll always have upset users during peak times because of call quality and dropped calls. The other is to set a quality level and guarantee that level for subscribers. CDMA can handle more simultaneous users than TDMA, but the question is how many? I've read everything from 2 to 15 times the capacity of TDMA. Many things suggest that in practice, it is closer to two times given a certain quality level.

CDMA has a variety of paths to get it from 2G to 3G technology as well. Two of the buzzwords you will hear related to CDMA's evolution are

IS-95B. (2.5G technology) IS-95A is the current version of CDMA most people in the United States use for data transfer at 14.4 kbps. Some locations will see an upgrade to IS-95B, which supports data rates of around 64 kbps.

1XRTT or CDMA 1x. Also called 1X (2.5G technology). 1X will increase data speeds from 14.4 to over 150 kbps.

3G, 4G, Oh Geez

Although 3G is not an endpoint (companies, organizations, and committees are already looking toward 4G), it could deliver what most people want: high-speed wireless data coverage. As with many of the other technologies we've covered in this chapter, 3G doesn't have a precise definition. However, one that is commonly used varies the data performance with what you are doing. The faster you move, the less throughput you get. While moving in a vehicle, you will see as little as 128 kbps and if the transceiver is in a fixed location expect 2 Mbps under *perfect* conditions. Figure 6.6 shows a comparison between the different 3G

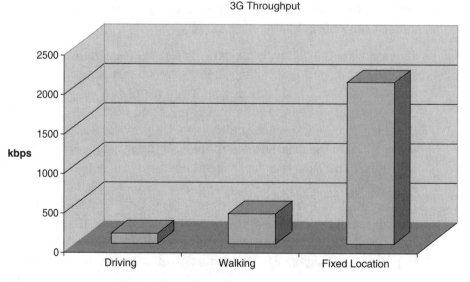

Figure 6.6 Ideal 3G speeds under different conditions.

speeds. Again we'll see a difference in theory versus practice in terms of what end users experience.

Unfortunately, 3G is not a single standard. There are two basic 3G standards in our future. Which one is dependent on the technology you use today. These include

WCDMA. Wideband CDMA (WCDMA) is the 3G endgame for GSM/TDMA carriers, which means Europe and potentially parts of the United States. That's right: At some point GSM/TDMA network operators will be switching to Wideband CDMA. On the positive side, it looks like a single standard (CDMA) will win in the end. Standards, in general, make the world of computing and communication less expensive for all.

CDMA2000. Existing CDMA networks will evolve to CDMA2000. This standard accomplishes 2 Mbps of throughput like WCDMA while using less of the spectrum for one air channel. Japan and United States will adopt the WCDMA standard.

The challenges of 3G are intense. Japan, the first country to rollout 3G (CDMA2000) and current users of 2.5G technology, delayed their introduction of 3G until Fall 2001 and reduced the scope of users. The electronics of both the network towers and handsets are complicated, and it will be some time until the bugs are worked out. Because of this, many believe that 2.5G will be the next big step for all, and whether the world goes to 3G standards is still tentative.

Either way, as a business, don't expect 3G technology any time soon. Also, don't wait for it to begin the rollout of wireless and handheld solutions for your customers and employees. Many large and small corporations are embracing wireless today as a 2G standard and shortly as a 2.5G standard. You'll be able to accomplish amazing amounts of business with these technologies alone.

WLAN Technologies

As I am sure you gather, the WWAN landscape is an absolute mess of standards. However, doesn't having many standards imply that you have no standard? I digress. As mentioned previously, WWAN technologies for speed, interference, and coverage reasons are not suitable for the interior of corporate facilities. Enter Wireless Local Area Network or WLAN technologies. They operate at speeds hundreds of times faster than WWAN wireless. For indoor settings or a corporate-wide campus, there are two well-recognized standards for wireless communication. Fortunately, they both have purposes within an enterprise and droves of devices are being developed with 801.11b and Bluetooth support. Let's look briefly at each and how they will affect your business.

802.11 (Wi-Fi)

802.11b is an IEEE standard for wireless Ethernet and is also called Wi-Fi. Many companies today are using it to allow workers to roam within a facility and have a connection to the corporate intranet or Internet from a laptop or desktop computer. Figure 6.7 shows how a company could utilize Wi-Fi technology within their business. The current standard operates in the 2.4 GHz spectrum (along with Bluetooth) and supports speeds of 2 through 11 Mbps. Setting the device to communicate

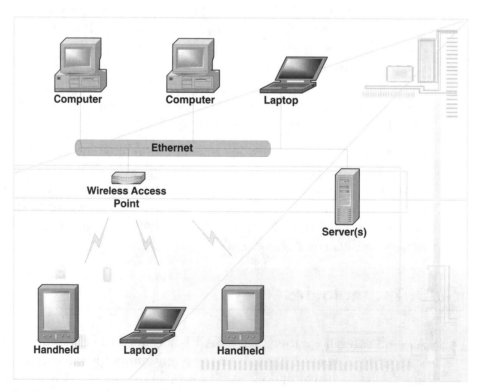

Figure 6.7 A wireless LAN architecture.

at a lower speed usually results in a wider coverage area. The upcoming 802.11a standard will support speeds in excess of 50 Mbps. In contrast, most regular, wired corporate networks run at 100 Mbps. Either technology has the performance to allow workers to conduct business anywhere on a campus. The signal penetrates walls enabling for a minimal number of access points to cover a floor of a building. A variety of companies sell 802.11b equipment. See Table 6.2 for a subset of these.

Users need only insert an 802.11 network card in their laptop and they are connected to any installed wireless access point. Figure 6.8 shows a wireless access point from Symbol technologies. This device is plugged

Table 6.2 802.11b Hardware Manufacturers

COMPANY
3Com
Apple
Cisco
Compaq
Lucent
RadioLAN
Xircom

Figure 6.8 A Symbol Spectrum 24 wireless access point.

into an Ethernet LAN and can create a wireless Ethernet connection between it and devices with 802.11b cards. The technology works well and is in place at many enterprises. Depending on the equipment used, power of the transceiver used, and interference from the building itself, one wireless access point can cover thousands of square feet and penetrate some types of walls. Costs are coming down quickly, which will enable new buildings to be built from the ground up with wireless infrastructure for less cost than a traditional wired network.

Wireless indoor networks, whether 802.11b or Bluetooth (described in the following), will open up all types of applications for handheld devices. Workers can walk around warehouses collecting inventory information and help desk technicians can take trouble ticket information with them throughout the buildings they support. Laptop users can do work anywhere inside the building, no longer limited by a network jack in a wall.

A number of companies have created ways to attach Wi-Fi cards to handheld devices (see Table 6.2). They work well with the exception of battery life. 802.11b standards assumed the computing devices using it would have a laptop size battery or be plugged into a power supply. So, 802.11 technology may have limited success with handheld devices. What is likely to supplant it will be Bluetooth.

Bluetooth

Bluetooth, considered a piconet technology, is a relatively new standard. About the time this book is published, a fair number of Bluetooth devices should be on the market supporting a number of needs. Bluetooth was developed to create a wireless standard to reduce the number of computer cables we use to connect devices and to enable small, low-power computing devices to communicate wirelessly. Its design includes features like inexpensiveness, low-power requirements, and automatic negotiation of connections. In one practical example, you will be able to move a Bluetooth-enabled Palm device within range (30 feet or less) of a desktop computer and have them immediately begin synchronizing. Obviously, security would need to be set up to prevent unwanted data loss. Other examples include phone headsets that work wirelessly with phones and computer keyboards that talk wirelessly to the computer.

If you look around on the Web, you will find no shortage of articles and columns discussing Bluetooth and its relationship to 802.11b. Some believe one will replace the other. Others believe one technology is inferior to the other. I believe that the two technologies will coexist. Bluetooth's sweet spot is wireless communication for small, low-cost, low-power devices. The technology will eventually become standard in cell phones, handheld devices, home appliances, and some PCs. Because both 802.11b and Bluetooth share the same frequency spectrum, there is bound to be some interference between the technologies. I believe as 802.11b makes way for the faster 802.11a standard (54 versus 11 Mbps) operating at a higher frequency, this interference will go away. General-purpose computers will use 802.11a networks for wireless communication while Bluetooth is used in the types of devices described previously.

A range of well-funded companies are out there designing new and interesting ways to have new devices talk with one another because Bluetooth supports the ad hoc setup of wireless connections between devices. The goal is to create better information flow, reduce costs, and create new revenue opportunities. Imagine a day where you walk into a shopping mall or an amusement park and the Bluetooth infrastructure sends a map of premises, information, and descriptions to your device all without your intervention, but with your permission.

As mentioned previously, Bluetooth trades range and bandwidth for long battery life. Bluetooth has a range of 30 feet from a small, battery-powered device. Performance is between 57.6 and 720 kbps depending on several factors, making it several times slower than 802.11b. Bluetooth also supports 128-bit encryption to prevent data eavesdropping. About the time this book is published, Bluetooth hardware and applications for handheld devices should be hitting the market.

Other Technologies

Location-based services (LBSs) is an emerging buzz phrase. It's not a technology as such. LBSs utilize existing wireless technology to allow a person to discover what services are physically near them. For example, you are wandering around downtown Chicago and you'd like to know what the nearest pizza place is to your location. With a wireless device, you'd enter the cross street at which you are standing, enter a zip code, or rely on the network to identify your location based on the antenna you're communicating with. Using an Internet database and this location information, the system would respond with the answer.

Some of the location-based services companies are piloting include

- Walking directions
- Driving directions
- Find the nearest movie theater, Kinkos, Starbucks, and so on
- 411
- E911
- Digital coupons for restaurants or stores nearby

The Kelsey Group predicts that by 2005, some $11 billion in revenue will be attributed to LBSs. There is no doubt that information like this is invaluable. How many times have you needed a particular service, but didn't know how to find the closest one? One of the largest challenges to the adoption of LBSs is the precision to which you can know your location. Imagine a wireless antenna that your cellular phone/handheld connects to. Because the location-based service knows the antenna you're connected to, it can tell you about any services you need within its range. Let's say the antenna has a range of five miles in any direction. That means you could be anywhere within a 75-square-mile range. That's not so accurate. What is needed is a more accurate way to alert these services of your location. Enter GPS.

The Global Positioning System (GPS) is another key technology we may take for granted in the coming years. GPS is a series of 24 satellites orbiting the earth. Because the satellites are in known locations, if you can figure out the distance to four of them, you know your exact location on earth. A GPS receiver, like the one in Figure 6.4, performs these calculations. There are many types. Some are standalone units. Others connect to Palm and Windows CE devices, and still others are mounted in cars. Depending on the system you use, you can know your location to within a few meters. With this type of precision, location-based services could take off. Over the past few years, the size of GPS receivers has decreased dramatically enabling for them to be unobtrusive. It may not be long until they are incorporated into cellular phones or handheld devices as standard equipment.

GPS, combined with location-based services, will create all types of new business opportunities, whether consumer or corporate. For example, GPS and location based services could be rolled out to provide trucking fleets with all types of information, including preferred places to fill up with gas and have maintenance done on the truck. GPS with a wireless connection can accurately track any vehicle for a customer or reroute truck dynamically.

The Bridge from Here to There

This chapter has covered a multitude of wireless technologies. We covered some of the basics related to wireless transmissions and showed how wireless data communication is in many ways similar to other wireless technologies like radio. 1G, 2G, and 3G were discussed to paint a picture of the past, present, and future of cellular technology. Alternatives to cellular data from various paging networks were covered. Additionally, the basics of CDMA, TDMA, and CDMA were compared. Table 6.3 recaps the speeds of these networks to give you a high level picture of speed versus a standard modem. The last entry in the

Table 6.3 Network Speeds for Comparison

TECHNOLOGY	SPEED (THEORETICAL)	COMPARISON TO 56K MODEM	TECHNOLOGY TYPE
Mobitex	8 kbps	.14x	WWAN
Re/Flex	9.6 kbps	.17x	WWAN
CDMA	14.4 kbps	.25x	WWAN
GSM	14.4 kbps	.25x	WWAN
CDPD	19.2 kbps	.33x	WWAN
2.5G	50–384 kbps	.87x–6.67x	WWAN
Bluetooth	56–720 kbps	1x–12.5x	WLAN
3G	128 kbps–2 Mbps	2.2x–35x	WWAN
802.11b	2–11 Mbps	35x–196x	WLAN
802.11a	54 Mbps	960x	WLAN
100 Base-T Ethernet (wired corporate LAN)	100 Mbps	1778x	Wired LAN

table covers a standard wired Ethernet LAN you would expect at any enterprise.

You may be wondering: Does someone make a handheld device or smart phone with both WWAN and WLAN technology built-in to handle workers who need coverage indoors and out? Today, no. However, by the time this book is published, there will likely be a number of announcements. All devices with native wireless have one or the other built-in. For those companies that require both types of coverage, you would need to purchase a device that has an expansion slot and two wireless cards, one for WLAN communication and one for WWAN communication.

Wireless networks and related technologies are complicated, acronym-related fields. However, many of them can be explained in layman's terms. This is important for enterprise because most true wireless solutions that cover thousands of workers across multiple geographies cannot be created with a single wireless network. Atlanta, for example, may use CDMA and GSM technology while Denver only has TDMA. (I'm not saying this is true. This is for the sake of an example.) Fortunately, enterprises can create one application that runs across multiple network types to support workers in these different geographies. This technology is covered in the next chapter on wireless middleware. Also, see the Resource section at the end of the book for a number of links to companies, papers, and articles on wireless technology that may be of interest to you.

Next it's on to one of the most exciting chapter of the book: wireless middleware. Middleware is the workhorse of enterprise solutions, whether wireless or not. It provides data synchronization to all types of back end systems, security, software updates, and hardware management to handheld devices.

CHAPTER 7

Wireless Middleware, Leprechauns, and a Magical Rainbow

Junior consultant: How does our clients' sales data get from their corporate database to their handheld devices?

Senior consultant: Our clients don't really want to know. As far as they're concerned, there is a leprechaun carrying their data in little black pots across a magical rainbow. They just want their data and applications to be accessible anytime and anywhere.

—Overheard during consultant training

So, perhaps the previous conversation didn't really happen, but the message is true: *Enterprise users want distributed access to their tools and data.* Over the last three decades, IT departments have fulfilled this need progressively better with systems based on terminal/host, client/server, and Internet/intranet technologies. (Rather than mentioning these three repeatedly, I'll refer to them as the *Big 3*.) At the time of their respective introductions, each of the Big 3 technologies granted evolutionary improvements in how users access and use their data and tools. These improvements are measured in terms of performance, user interface, depth and breadth of data, and robustness of toolsets. Wireless computing is the next revolutionary technology in this chain, and it delivers two benefits never before available.

The first benefit of wireless computing is mobility. As discussed in other chapters in this book, a handheld device shrinks a significant portion of the power and storage of a traditional desktop computer to fit in the palm of your hand. This shrinking is due to the progression of chip technology, display technology, and input/output improvements (like wireless, voice recognition, and so on). Similar to the Big 3, handheld devices enable IT departments to distribute computing power closer to the user's need via low-cost hardware and software. This happened in the 1970s with the distribution of terminals that accessed hosts, in the 1980s with the distribution of clients that accessed servers, and in the 1990s with the distribution of browsers that accessed Web servers. (Also of note is that the acceptance of handheld devices has followed the trail blazed by personal computers: their introduction with hobbyists, followed by wide acceptance by consumers, and finally being leveraged in the enterprise.)

The distribution of this computing power grants users greater control over how efficiently computing resources are used. The bottom line is easier access to appropriate tools at more critical times and locations. Who cares? Anyone who has needed directions while traveling, collected data on paper with a clipboard, or had to guess when answering a question in front of a customer. Handheld devices (and even portable computers) enable users to exploit computing power at more appropriate locations and crucial times. By itself mobility is useful, but to unlock the full potential of wireless computing relies on connectivity *to information*.

The second benefit of wireless computing is connectivity. Without connectivity, wireless computing would be a step backwards to independent computing resources and islands of data. Delivering connectivity to handheld devices is the natural progression from Ethernet and Internet technologies. Connectivity enables users to leverage existing computing resources while not being tethered to any particular location. This is the further distribution of computing resources, extending the progression started by the Big 3.

Merely enabling connectivity is not enough in real-world wireless systems. Business applications today run on a variety of platforms, are created and sold by hundreds of vendors, and support a variety of different standards. Implementation of business applications in an enterprise has always been fraught with challenges of integration. What is needed is a tool that will enable communication between applications speaking different languages. This tool is middleware.

If they haven't already, enterprise IT groups will be building and deploying systems for wireless computing, in spite of the fact that they generally have little or no experience with the technology. The purpose of this chapter is to explore wireless middleware in order to educate enterprise IT groups about its appropriate use. Most of the chapter will cover the architecture (why, who, how, what, when, and where) of wireless middleware. We will illustrate how its concepts are addressed within actual enterprise IT organizations and point out where special consideration is necessary to address common challenges. Finally we will discuss keys to the future of wireless middleware, including trends, evolving standards, and next generation developments.

Middleware Overview

By definition, middleware is translation software that enables applications to communicate regardless of platforms, standards, or vendors. All distributed computing technologies rely on middleware to enable communication, including the Big 3. Middleware enables terminals to submit transactions to hosts, clients to view data from a DBMS, and browsers to surf sites from Web servers.

Specifically, communication enabled by middleware is typically used to deliver one or more of the following services to an overall system:

Networking. This service is the most important one provided by middleware. A *network* is two or more computing nodes connected so that they can share information. Although middleware's networking service does not actually establish the connection, it does assist in managing the connection and translating any communication over the connection. For example, a user may want to look for a file on the corporate network, but they don't have to worry about the various vendors, operating systems, or file systems of the file servers. Middleware handles the translation between the PC and the file servers so that the user has a simple, seamless experience.

Message processing. Dubbed Message Oriented Middleware (MOM), this middleware service enables applications to communicate with one another indirectly by passing messages through a message service. This requires the applications to connect to just the MOM and not every application or device with which it wants to communicate. As an analogy, we could all deliver our own mail, but the post office does it much better (most of the time). Similar to the post office, MOM will take actions to guarantee delivery, such as message prioritization, verification of delivery, and message storage, until a recipient application is available.

Transaction processing. Middleware that provides this service is dubbed a TP monitor. This service ensures that transactions are processed in a reliable, timely manner with support for commit and rollback functions. For example, what appears to a system user as a single transaction for creating a requisition in an e-commerce system might actually require several actions to take place: checking prices, verifying availability, processing a credit card charge, and forwarding purchase orders to one or more vendors. The TP monitor manages communication with the variety of applications that perform individual actions such as online catalogs, real time inventory, charge processing, and vendor's ordering engines. It then commits the overall transaction only if all actions are complete; otherwise, it rolls back all actions in the transaction so as not to leave the transaction partially complete.

Database connectivity. This service enables applications to process data (request, aggregate, and publish/subscribe) via a DBMS-independent interface. The benefit is that applications don't have to know all the languages used by databases, just any industry standard database interfaces like Microsoft ODBC, Microsoft OLEDB, or Sun/Java JDBC.

As the Big 3 technologies have matured, traditional middleware has become transparent to users, which in turn has set expectations very high for middleware in wireless computing. The distance and ease with which wireless voice communications occur today raise the bar even higher. In delivering against these high user expectations, middleware for wireless computing faces additional new challenges. Availability, reliability, bandwidth, battery life, and a variety of devices and wireless data protocols all conspire to degrade the user experience of connectivity with wireless computing. This is where wireless middleware enters the picture.

Wireless middleware is a new breed of middleware specifically designed to meet the challenges of wireless computing. It provides services similar to traditional middleware as well as some additional services specifically designed for wireless computing. Generally speaking, four services are provided by wireless middleware:

Secure communication management. This service is similar to the networking service provided by traditional middleware: managing the connection between wireless device and back-end application, but with particular emphasis on secure communication. This wireless middleware service is designed to handle the harsh conditions inherent with wireless connections such as dropped connections, high latency, and low bandwidth.

Synchronization. This service examines two sets of data and performs actions to make them the same, or *synchronize* them. This service is particularly valuable in wireless systems where devices can be disconnected for long periods of time. When a connection is finally established, the data on the wireless device and data on the back-end application can be synchronized, ensuring that both data sets are up to date.

Message processing. Again, this service is similar to the message processing service provided by traditional middleware. However, traditional message processing services are typically put in place to resend messages because recipients are too busy to respond to the initial message. In wireless systems, message processing faces a different challenge in that the recipient cannot be reached because a connection is not present. Messages build up in queues on a wireless device while waiting for a connection to be re-established.

Management tools. IT resources face big challenges when introducing a whole new computing platform to their enterprise. At minimum they need to provide the same services to these new devices that they do to their existing computing resources: security, backup and restore, electronic software distribution, and asset management.

If you have a PocketPC or Palm handheld, you have already seen wireless middleware in action, even without a wireless connection. Devices with these operating systems ship with PC-based wireless middleware, Microsoft's ActiveSync for PocketPC devices, and Palm's HotSync Manager for Palm OS devices. These products manage communication between the devices and the PC through serial or USB connections and provide synchronization services for applications like email, contacts, task lists, and calendar. ActiveSync ships with an AvantGo client for synchronizing Web content for offline viewing as well.

If you have a wireless modem connected to one of these devices, you may have experienced message processing as well. Third party subscription services from vendors like OmniSky, GoAmerica, and Sprint actually host wireless middleware messaging services that process incoming and outgoing email, instant messaging, and text paging.

Why Is Wireless Middleware Necessary?

In truth, wireless systems can be implemented without wireless middleware (see Figure 7.1). They need to handle the services described previously with or without middleware. Consider an example company called MFG Inc. whose sales personnel visit clients, create orders for widgets on a wireless device, and submit them to a central database at the MFG Inc. main office.

The application must drive an attached modem, open a wireless network connection to the main office, authenticate against one or more back-end applications, create and transmit messages over the connection, receive and interpret messages from the database, adjust for network latency and dropped connections, and so on. All this complexity is to handle *communication functionality*, and we have not even talked about the *business functionality* that the application provides.

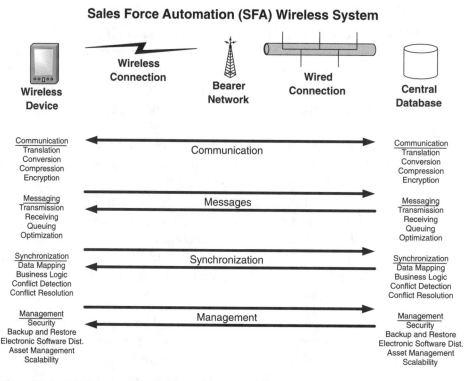

Figure 7.1 Wireless communication without middleware.

Moving these communication services out of the application and into middleware provides some benefits:

- Specialized, reusable middleware can focus on what it does best: communication. Other components of a system, such as the wireless ordering application and the central database of my example, can focus on what they do best: business.

- Monitoring and supporting middleware that is centralized on an enterprise server is easier than for those same services deployed to every component of the system. Back to the wireless ordering example, imagine the difficulty in troubleshooting a communication problem on dozens of devices out in the field instead of on a central server.

In addition to the complexity involved with communication, wireless systems face some new challenges never before seen in the wired world. Wireless middleware is specifically designed to overcome these challenges:

- Developers face a variety of complex network protocols and diverse device platforms, sometimes all within a single wireless system. Wireless middleware provides a simple programming interface against which developers can write code, thereby insulating them from the network and device differences. From the developer's perspective, the communication seems network and device independent. In general, wireless middleware enables developers to focus on application functionality rather than the *plumbing* of connectivity and communication.

- Wireless provides low bandwidth connections. Wireless middleware can include compression and optimization features to maximize the throughput of a wireless connection.

- Conditions for wireless connections are never ideal. Connections may be difficult to establish due to coverage limitations and physical impediments like buildings or mountains. Once established, connections are still at risk due to device or weather. Note that these risks apply to all types of wireless connections, from piconet to LAN to WAN. Wireless middleware is designed for the unpredictable nature of wireless connections. It can adjust the packet sizes of data, relax timers, cache sessions until reconnection, or even switch to another network in order to preserve communication.

- Communication through the air has inherent security risks. Wireless middleware can combine authentication and encryption techniques to provide secure communication.

- Wireless devices will eventually have a period when they are not connected, but are still being used. Although disconnected, perhaps transactions were cached on one side, or data was updated, or messages queued. When a connection is re-established, the changes both on the wireless device and the back end need to be reconciled. Wireless middleware can include the engine for storage and forwarding of changes as well as business logic for conflict resolution.

- Users carry wireless devices with them at all times and frequently consider them as personal items. Users come to expect a very personalized experience when using wireless applications. Wireless middleware can help applications to personalize the user experience by providing access to the user's subscriber number, device identification, and location.

All of these wireless system challenges could be overcome with custom-programmed functions within the application itself. However, wireless middleware, as a separate component in a wireless system, encapsulates the functions necessary to enable communication. It enables each of the other components, or *actors*, in a wireless system to focus on what it does best. So, to summarize the past few pages: Wireless middleware is good. Purchasing it is much less expensive than developing this communications infrastructure yourself (many times over).

Who Are the Actors in Wireless Computing?

We've introduced our example company, MFG Inc., with their sales personnel submitting orders from a wireless device, and we will continue to expand this example throughout the chapter. I've been designing IT systems for almost 10 years, and I've learned that a good way to formalize a process is by writing a *Use Case* that describes exactly how the process goals are actually attained in a system. A Use Case, if you are unfamiliar with the term, tells a story of how an actor in the system accomplishes his or her goal step-by-step. It also handles the case of what happens if any one of the steps fail. Use Cases are excellent for gathering requirements from clients and helping to flesh out the design of systems. There is no wrong way to write a Use Case, just adopt a consistent format that suits your needs. I recommend *Writing Effective Use Cases* by Alistair Cockburn (Addison-Wesley, ISBN: 0201702258) for a very good treatment of the subject.

Furthermore, the Use Case helps us clearly identify the components (actors) in a system. This section will talk about the actors in a wireless system and how they interact in general and specifically with the wireless middleware (see Figure 7.2.)

For MFG Inc., the goal for their sales personnel (and, therefore, the Use Case) performing the ordering process within the sales force automation system is to *submit the order to the central database*. There may be other employees performing other processes and using other systems at MFG Inc., but this Use Case is focused on our example situation. To achieve this goal, actors perform a sequence of actions. Actors are simply components of the system, including people, organizations, devices, or applications. In most systems, the actors typically consist of a user, one or more applications wanting to communicate, and aptly named middleware helping the applications to communicate. In the previous example, the actors are the salesperson, wireless device, ordering application, central database, and wireless middleware. Each application can be a sender of information or a receiver, or alternate as both. The middleware's job is to manage the communication between the applications so that each application performs the actions required to complete a Use Case.

When we start to talk specifically about wireless computing, the actors can include wireless application users, wireless devices, wireless applications, back-end applications, and wireless middleware that moderate all communication.

Wireless Application Users

Wireless application users can literally be anyone with a relationship to your enterprise who does not have physical access to its stationary computing resources (see Figure 7.3).

They are the ones who will use the wireless application to achieve a certain goal, maybe for themselves, their organization, or their enterprise. In our ordering example for MFG Inc., the wireless application

Use Case: Submit Order to Central Database

Actors: salesperson, wireless device, ordering application, central database, wireless middleware

Scope: Sales Force Automation (SFA) wireless system

Precondition: One or more orders created in the ordering application on the wireless device

Minimal Guarantee: Attempt to submit order is logged in the ordering application

Success Guarantee: Order is submitted to the central database

Trigger: Upon establishing connection between wireless device and central database

Main Success Scenario

1. Ordering application requests a session with the central database.
2. Wireless middleware opens session.
3. Ordering application sends order data via the session.
4. Wireless middleware converts the order data.
5. Wireless middleware writes the order data to the central database.
6. Wireless middleware confirms order data submitted to central database.
7. Ordering application receives confirmation of order data submittal.
8. Ordering application requests closing the session.
9. Wireless middleware closes session.

Extensions

3a. Connection is lost while submitting order.
   3a1. Wireless middleware re-establishes connection.
5a. Session is terminated while writing order data.
   4a1. Wireless middleware re-establishes session.
6a. Connection is lost while confirming the order data submittal.
   5a1. Wireless middleware places confirmation message in queue.
   5a2. Confirmation message delivered upon re-establishing connection.
7a. Connection is lost while requesting close of session.
   7a1. Wireless middleware closes session after time-out period.

Figure 7.2 Use Case for ordering.

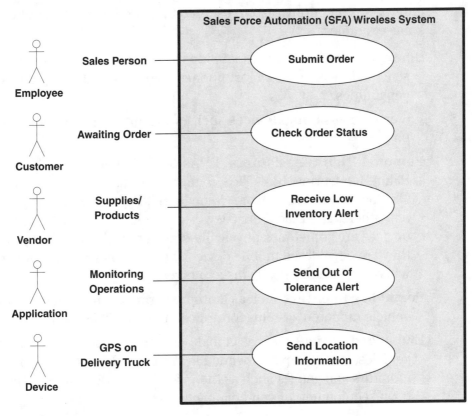

Figure 7.3 Illustration of wireless application users.

users are employees, specifically sales personnel. In general, they can be employees, customers, or vendors. In a more abstract sense, they can even be other applications or devices. When designing a wireless system, it is crucial to consider all possible wireless application users and design for the appropriate subset so as to maximize your investment. Why build a system for a small group of users when others could also benefit?

Wireless Devices

Mobile devices are broadly defined as devices that perform distributed computing and are occasionally physically disconnected from other enterprise computing resources. A complicated definition, but here are some examples:

Smart phones that have capabilities to run wireless applications and send and receive data. Most phones manufactured today have these capabilities.

Pager or messaging devices such as a traditional numeric or text pager.

Personal Digital Assistants (PDAs) such as Palm OS devices from Palm, Handspring, IBM, TRG, Symbol, Sony, and others; Windows CE/PocketPC devices from Compaq, HP, Fujitsu, Hitachi, Symbol, and others; and Blackberry devices from RIM. Although these devices are sometimes physically connected via a cradle or modem, they are considered mobile devices because they can still function when they are physically disconnected.

Portable computers such as laptops, portable barcode scanners, or vehicle mounted systems for police, fire, and EMS.

Embedded devices such as global positioning system (GPS) devices in vehicles, smart chips embedded in credit cards, or computerized monitors and gauges such as those found in utility meters, manufacturing equipment, or vending machines.

At MFG Inc., the sales personnel will be using PDAs to create and submit orders, but other devices could be used as well. Customers might use smart phones to quickly check order status; service personnel might be dispatched by service request messages sent to their pagers; executive management might monitor operations via software on their laptop; and manufacturing equipment might update a database with quality control measurements.

Wireless Applications

This one is obvious: Wireless applications are tools running on a wireless device. Some examples are also obvious as most wireless devices ship with one or more of them:

Messaging such as email, text, and numeric paging, and instant messaging

Personal Information Manager (PIM) for managing the user's address book, schedule, notes, and other personal information

Web/WAP browser for viewing HTML/WML content

Entertainment such as gaming, music, and video applications

Reference materials that vary from device instructions, to dictionaries, to recipes

These applications are mostly for consumer use and, as such, will communicate with other applications that consumers use, such as Web sites, their MP3 files on their PC, or desktop applications like Microsoft Outlook. However, a whole other class of back-end applications run on wireless devices that are targeted at enterprise users rather than consumer users:

Executive Information Systems (EIS) to provide corporate management visibility into operation performance metrics

Enterprise communication to enable people to stay tied to enterprise communication resources such as email, discussion groups, and alerts

Groupware, collaboration, and knowledgebase to enable groups to electronically share information relevant to their everyday work

Sales Force Automation (SFA) to manage client lists, sales campaigns, commission tracking, and so on

Customer Relationship Management (CRM) to provide employees and customers visibility into operations including order processing and tracking, invoice review and payment, and customer support

Enterprise Resource Planning (ERP) to track the supply chain activities, raw materials, work in process, and finished goods

Field Force Automation (FFA) to electronically assist field workers in data dissemination, collection, and transmittal

Dispatch and location to manage the work orders, routes, and current status and location of mobile resources such as field workers and assets

In our MFG Inc. example, we will talk mostly about SFA, but the concepts in this chapter apply to any type of wireless application in an enterprise.

Back-End Applications

The previous examples of wireless applications targeted at enterprise users should give you an indication of the back-end applications that could be involved in a wireless system. The list could extend to every enterprise application you currently distribute to your internal employees. It could further extend to applications you distribute to your customers, vendors, partners, and investors. It could even conceivably extend to every process your enterprise has, regardless of whether it has already been automated with computing technology.

To help identify back-end applications in your own enterprise, here are some examples of the technology on which the back-end applications run:

Host applications can communicate with wireless devices employing screen scraping techniques or running a terminal emulator.

Server applications can communicate with wireless devices running application clients.

Intranet/extranet resources or applications can communicate with wireless devices running a browser.

Database management systems can communicate with a wireless device through an API or generically through ODBC, OLEDB, or JDBC.

Most back-end systems contain one or more of these technologies. Consider Figure 7.4 to determine which are considered direct actors in a wireless system and which are indirect.

The illustration shows that the only back-end applications to be concerned with in a wireless system are those that wireless middleware will directly access. For our MFG Inc. example, an internal ordering system may have a database to store data, an application server to house business logic, and a Web server to provide the user interface. The wireless middleware may submit transactions to the application server, or through the Web interface.

Wireless Middleware

According to the definition, wireless middleware is everything between the wireless applications and the back-end applications whose purpose

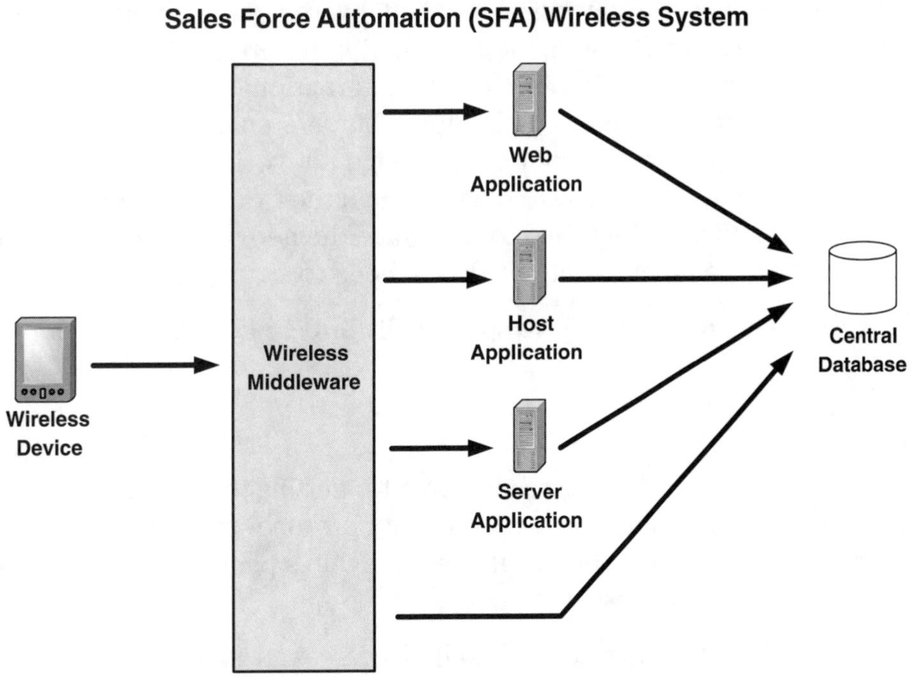

Sales Force Automation (SFA) Wireless System

Figure 7.4   Illustration of back-end application actors in a wireless system.

is to move and/or transform data to enable communication. It is the glue or the plumbing that connects all the pieces. In practice, the lines often blur between wireless applications, back-end applications, and wireless middleware. Is a database of usernames and passwords for authenticating considered to be middleware or part of the back-end application? Is a message queue that batches up transactions on the wireless device considered to be middleware or part of the wireless application? Although no clear delineations exist, the core service that is always provided by wireless middleware is communication management.

How Does Wireless Middleware Manage Communication?

It is the responsibility of the wireless middleware to manage communication so as to optimize the overall wireless system operation. To frame a discussion of how wireless middleware manages communication, I'll use the Open Systems Interconnection Reference Model (OSI-RM) to set the stage (see Table 7.1). The International Organization for Standardization (ISO) has maintained this reference model since 1984 as an abstract description of how to design networking protocols divided into seven layers. (Note that although the OSI is an international standard and is widely used to introduce the networking concepts, actual network implementations do not need to strictly adhere to it.)

Let's examine the layers specifically affected by wireless middleware.

Network Layer

The Network layer is responsible for getting the data from one network node to another. The most common protocol used in the Network layer is the Internet Protocol (IP), which routes packets of data that are labeled with a target address.

Wireless devices often switch IP addresses as connections are dropped and reopened, as they roam between subnets in a wireless network, or even as they switch from one network to another. Wireless middleware can manage the device's IP address as it is visible to the network in order to make these wireless challenges transparent to the user (see Figure 7.5).

Table 7.1 Open Systems Interconnection Reference Model

LAYER	DESCRIPTION
1. Physical	Converts data to the corresponding signals transmitted over a communications channel
2. Data Link	Defines the logical organization of bits to be transmitted
3. Network (*)	Delivers data between network nodes using routing and optimization functions
4. Transport (*)	Defines the quality and nature of the data delivery
5. Session (*)	Provides the mechanism for managing the dialogue between applications
6. Presentation (*)	Describes the syntax of data being transferred
7. Application (*)	Performs common application services

Source: International Organization for Standardization, OSI 7498
(*) indicates a layer affected by wireless middleware.

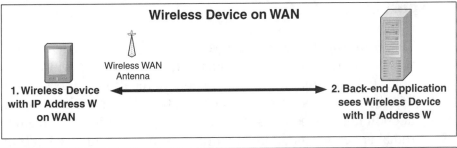

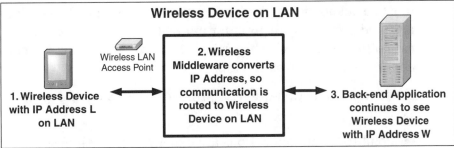

Figure 7.5 Illustration of IP address management in the Network layer.

Consider a shipping application running on a wireless device in a vehicle in one of MFG Inc.'s delivery trucks. During the vehicle's delivery route, the application sends location and delivery confirmation updates to the warehouse's central database via a wireless WAN. Although wireless WAN communication is slow, it is the only appropriate technology while the vehicle is more than 300 feet away from the warehouse. While communicating on the wireless WAN, the wireless device in the truck is identified to the rest of the network with a certain IP address.

The warehouse dock at MFG Inc. is equipped with a wireless LAN. When the vehicle returns to the warehouse dock, it begins communicating over the wireless LAN because it is cheaper and faster than its current wireless WAN connection. The challenge is that the IP address of the device is recognized on the wireless WAN and not on the wireless LAN. Transparent to the user, the wireless middleware steps in to negotiate a new IP address for the device on the wireless LAN and transitions the connection from one network to another. With this new connection on the faster wireless LAN, the device begins to load information for the next delivery route from the central database.

Transport Layer

The *Transport layer* is responsible for managing the quality and priority of data transmittals. The most common is Transport Control Protocol (TCP) for reliable end-to-end communication in the wired world. TCP is connection oriented, meaning that a logical, point-to-point connection must be established before communication can begin and must remain open until communication is completed. Also, each transmittal must be acknowledged before another can be sent. Connection-oriented protocols are great for interactive applications that communicate often, such as Web browsing. Once the connection is open, communication is fast and reliable.

However, connection-oriented protocols pose a challenge in a wireless system:

- Point-to-point connections are expensive in a wireless WAN. Carriers charge more to establish point-to-point connections because they must open and maintain a communication channel dedicated to the connection. Carriers have a limited supply of channels and must allocate them between data communications and higher revenue voice communications.

- Connections can drop in a wireless environment due to environmental factors like weather or interference. A connection cannot be recovered once it is dropped, requiring a new connection to be opened.

- Connection-oriented protocols are greatly impacted by network latency. Each transmittal must be sent, received, and acknowledged before another one can be sent. If the latency during a send/receive/acknowledge cycle takes milliseconds on a wired network, it could take a full second or more on a wireless network. Sending hundreds of transmittals on that same wireless network, as would be required for a Web page or email attachment, could conceivably take up to a minute or more. An analogy for this is delivering product from our MFG Inc. warehouse to a customer location. Consider a highway (connection) that is closed to all other traffic and a delivery truck taking one package (transmittal) at a time between locations, with the travel time in between representing the network latency. The latency would be low if the warehouse is around the corner from the customer, but it would be high if 1,000 miles separate them. High latency will make the entire moving process quite long for 100 packages, and therefore 100 round trips.

Wireless middleware can overcome these challenges in two ways. First, it can adapt on-the-fly to harsh wireless conditions by adjusting connection parameters such as transmittal sizes, time-out tolerances and retry limits. In our delivery analogy, this is the equivalent of changing the amount of packages taken on each delivery trip, the time a truck is allowed to be away from the warehouse, and how often the truck is sent to the customer. Wireless middleware can change these parameters so as to optimize communication.

Alternatives to connection-oriented protocols at the Transport layer are called connectionless protocols, such as User Datagram Protocol (UDP) or Short Messaging Service (SMS). These protocols do not need to establish a point-to-point connection, instead they use a channel that is always open and shared by all transmittals on the network. Carriers have to set aside less bandwidth for this single shared connection than several private connections, meaning lower costs that are passed down to the users. In our MFG Inc. delivery example, this is the equivalent of using a common highway for trucks from many companies delivering mail to many customers. In addition, connectionless protocols can transmit data without waiting for confirmation of its receipt, like a warehouse dispatching another truck without waiting to confirm that previous delivery was successful. This means that the effects of network latency are greatly reduced because transmittals are sent asynchronously, or independent of one another. Taken together, the open channel and asynchronous transmittal features combine to make connectionless protocols significantly faster in a wireless system, but at the cost of reliability. Wireless middleware can leverage a connectionless protocol for its speed and can employ its own measures to ensure reliable communication.

Session Layer

Where the Transport layer manages the connection between two network nodes, the *Session layer* manages the communication between two applications over that connection. Going back to our MFG Inc. ordering example, once the application establishes a connection to the main office, it then must communicate with the central database by establishing a session. To establish the session, the wireless application may send information about the user, the application, and the device. The session could then verify the user's identity, set preferences based on the user's profile, and even configure functionality based on user preferences and the types of device and connection. All of this happens once at the beginning of the session and is persisted throughout the lifetime of the session.

If you have ever connected to an Internet Service Provider (ISP) and had to provide log-in information, then you have experienced the establishment of a session. If you have ever lost a connection and been kicked off a system, you have also experienced the frustration of having to restart a session. At minimum you had to wait for the session to be restarted. Perhaps you were in the middle of a task and lost your work. Or worse yet, you were submitting a credit card purchase and you don't know if it was completed before you were kicked off.

Because wireless systems suffer from dropped connections, they in turn also suffer from prematurely terminated sessions. Wireless middleware can manage sessions in such a way as to enable them to continue across successive connections. Not only does that avoid the time consuming login and session restart actions, but it also prevents lost work and unconfirmed transactions.

Presentation Layer

Although this layer's *presentation* moniker might imply that it converts content for display on different device platforms, it does not. The *Presentation layer* is responsible for handling data conversion, compression, and encryption between applications. Presentation refers to how data is viewed by each application, not each user. Although wireless middleware functions at the Presentation layer are crucial to the performance and security of wireless systems, they are completely transparent to the users.

Data conversion. An unexciting, but accurate example of data conversion performed by middleware is the conversion of floating point numbers between applications with different math formats. The data conversions performed by the Presentation layer are syntactical, not related to a user interface. In fact, data standards such as ASCII (text), JPEG (graphics format), and MPEG (video format) have significantly reduced the need for conversions historically performed at the Presentation layer.

Data compression. Middleware compresses data before its transmittal and decompresses after its receipt. A wireless connection can have significantly less bandwidth than a similar wired connection. The compression performed by wireless middleware enables more data to be sent over this bandwidth more efficiently, thereby maximizing the performance of the connection.

Data encryption. In the Presentation layer, data encryption is just another form of conversion. Before transmittal, data is converted to an encrypted form, and after receipt, it is converted to a decrypted form. Wireless middleware performs the data encryption using industry standard encryption algorithms like Elliptical Curve Cryptography (ECC) and Triple DES.

Application Layer

The *Application layer* acts on communication requests made by an application. It is with this layer that the application actually interacts when it wants to communicate. Common protocols at the Application layer include Hyper Text Transfer Protocol (HTTP), File Transfer Protocol (FTP), Simple Mail Transfer Protocol (SMTP), and Telnet. Middleware can listen for requests from applications in these specific protocols, or languages. The middleware then takes over and uses functions at lower layers in the OSI Reference Model to manage communication.

It is at the Application layer that wireless middleware provides two additional benefits: personalization and content transcoding.

Personalization

A traditional system may know little more about its user than their login ID and preferences, and therefore is only able to provide a generic user experience. In general, users have different expectations of their user experience on wireless systems. Wireless middleware is capable of enriching the user experience by personalizing it. Wireless middleware can capture details about the connection speed, wireless device model, and physical location, and then feed them to the backend application. Imagine a personalized user experience for a MFG Inc.

salesperson who is browsing the sales lead database from the company's intranet site while at a coffee shop:

- Wireless middleware has detected that he is no longer working from his desktop computer and will forward message notifications to his device for the duration of the session.

- Wireless middleware detects the model of his device and knows that it has a built-in jog dial (a small dial on the side of a device controlled by your thumb and meant for scrolling), so the content enables its use in navigation.

- He has configured his system preferences to note the industries that he prefers to sell in. Now that he has logged in to a session with the system, wireless middleware provides profile information to the application so that he is provided with updated leads in his chosen industries prioritized first.

- Wireless middleware uses information about his connection to determine his location to within a few blocks. Because he is browsing the sales lead database, the system provides the list of leads prioritized by proximity to the coffee shop.

Content Transcoding

Several form factors exist for wireless devices, complicated by the variety of vendors, models, and browsers available. Using the previous personalization information, wireless middleware can deliver content in styles, formats, and languages appropriate to the user's device. As a user requests data, the source is scanned and reformatted in real time, then sent to the user. Let's continue the previous example of our MFG Inc. salesperson browsing through sales leads from a wireless device:

- Wireless middleware knows the dimensions of his screen so it reformats the content to avoid horizontal scrolling.

- Wireless middleware knows that his device has a monochrome screen, so it converts all graphic content to the appropriate supported grayscale depth.

- Wireless middleware knows that his device's micro-browser does not support frames, scripting, and various other advanced browser functionality, so it strips these out of the content before sending it to the device.

Putting It All Together

Let's look from the user's perspective at how wireless middleware uses all these functions at various layers to enhance communication. We'll look at a sample communication that might occur in our previous wireless Web-browsing example (see Figure 7.6).

- The back-end Web server provides Web content in response to the user's click on a link to the sales lead database on the company's intranet.

- Application layer: The Web content is reformatted for the user's device, the lead list is prioritized by proximity to the coffee shop, and the content is packaged as an HTTP response message so that the micro-browser on the device can understand it.

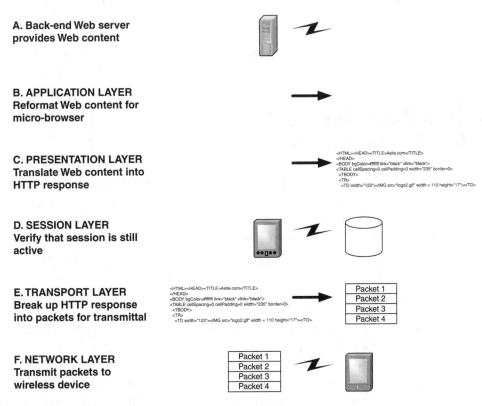

A. Back-end Web server provides Web content

B. APPLICATION LAYER Reformat Web content for micro-browser

C. PRESENTATION LAYER Translate Web content into HTTP response

D. SESSION LAYER Verify that session is still active

E. TRANSPORT LAYER Break up HTTP response into packets for transmittal

F. NETWORK LAYER Transmit packets to wireless device

Figure 7.6   User's perspective of wireless middleware communication service.

- Presentation layer: The HTTP response message is compressed so that it will use less bandwidth over a wireless connection. It will be decompressed once it reaches the device.

- Session layer: The user's session is verified to ensure that he is still logged in since the original request. If a connection had been dropped and reconnected, the wireless middleware preserves the session across the two connections.

- Transport layer: The compressed HTTP response message is broken up into packets for transmittal. Wireless middleware decides the appropriate packet size and other configuration parameters based on device and connection information. The packets will be reassembled once they are received on the device.

- Network layer: The packets are routed to the IP address of the device. If the device has moved to another network, say by returning to the office wireless LAN, then the wireless middleware will manage finding the device on that other network and routing the packets to it.

When communication is managed correctly by middleware, the fact that the user is browsing from a wireless device or his PC is transparent.

What Other Services Does Wireless Middleware Provide?

In addition to enabling communication, wireless middleware can provide any number of other services that contribute to a successful wireless system. These services could be built into the wireless or back-end applications, but their inclusion in wireless middleware increases the value of this middle tier.

Synchronization

One of the realities of wireless systems is that wireless applications eventually will operate while disconnected, often by design. Wireless middleware can ensure that disconnected wireless applications and data are brought up-to-date when a connection is established by synchronizing with a back-end application. This approach can be significantly cheaper than the alternative: maintaining communication so that changes can be pushed out from the back end and submitted from the wireless application.

Synchronization provided by wireless middleware can vary significantly in its implementation:

Data mapping enables the mapping of disparate data sources on the wireless application and on the back-end application. Data may take the form of records in tables, files in a file structure, or even Web pages on a Web server.

Native replication will synchronize data between databases from the same vendor on different devices (read: wireless device and back end). Native replication is more powerful than data mapping because it leverages features of the vendor's data engine for publication/ subscription and conflict resolution.

Custom programmed synchronization basically hands over the duties of synchronization to a custom-programmed engine. This approach provides the most flexibility, but also could require significant effort to develop.

What actually takes place during synchronization?

One-Way Overwrite

One-way overwrite is the simplest method of synchronization (see Figure 7.7). In this method, an entire data set from a source is copied over the old data set at a destination. For example, suppose at MFG Inc. that the sales department issued memos each morning regarding product availability, commission reports, and progress against sales targets. When a salesperson synchronizes his device with the back-end SFA application, he would receive the current day's memos, which would overwrite the ones from the previous day.

Because it is the simplest of synchronization methods, the one-way overwrite method has its limitations. For example, if a salesperson noticed that his commission report was wrong, he could not update it and synchronize it back to the back-end SFA application. Each synchronization would simply overwrite any changes he had made. Instead he would have to log on to the SFA application from a PC, make the changes to the central data set, and then resynchronize to give the changes back to the wireless device.

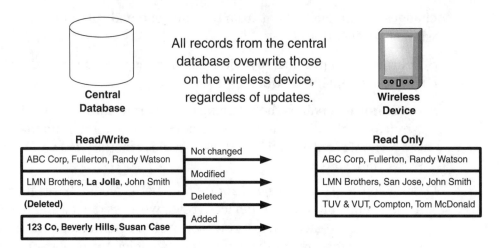

Figure 7.7 One-way overwrite method of synchronization.

This brings up the second limitation: All data is synchronized regardless of whether it has changed. If the data set is small, this might not matter. But if the data set is very large, say an entire product catalog or company directory, it might not make sense to give out an entire data set when little of it has changed since the last synchronization.

In general, the one-way overwrite method is appropriate for

- Data that has only one publisher and one or more subscribers, that is, only one data set is changed, and then pushed out to other recipients who cannot themselves change the data.

- Data where most or all of a data set changes at about the same frequency with which synchronizations occur. For example, all data is updated nightly and synchronized the following morning. If synchronizations occur more frequently, recipients will just be wasting time getting full copies of data sets that they already had.

One-Way Incremental

The one-way incremental method of synchronization overcomes two shortcomings of the one-way overwrite method, but at the cost of complexity to implement (see Figure 7.8). One-way incremental does not have the limitation of small data sets, nor do changes have to occur at the same frequency with which synchronizations occur. The third limitation of the one-way overwrite method is still present: Changes can only be made to one data set, and then given out to recipients.

With the one-way incremental method, only changes that have been made since the last synchronization are given out to recipients. For our SFA system at MFG Inc., their wireless middleware may synchronize product catalog updates out to each salesperson's device. Only changes that have been made to the catalog would be given to each recipient, greatly reducing the amount of data passed during each synchronization. In addition, the salesperson would not be concerned with how frequently he synchronized, as he would be with the one-way overwrite method. Because only changes are given to each recipient, there is no risk of wasting time getting the same data that is already on the device. If no changes have occurred, then no data is synchronized.

With any synchronization method that is incremental, there must be a method for detecting changes to data sets. If you can detect a change in a data set (that is, you can detect when data is new, updated, or deleted), then you can cut down on the amount of information you must send out. Changes can be tracked in three ways:

Record level tracking. This indicates that a record has been changed, regardless of how many fields on the record are affected. Record level changes do not track how many times the record has been changed, but instead just note the current values in the record. When

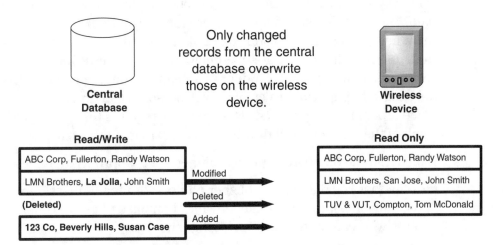

Figure 7.8 One-way incremental method of synchronization.

incremental synchronization occurs, the entire record is given to the recipient, even if only one field on the record has changed. An example is synchronizing an entire record for an address book entry when only the phone number has changed.

Field level tracking. This indicates the fact that a field within a record has been changed. Similar to record level tracking, this method does not track how many times the field has changed, but instead just notes the current value in the field. This type of tracking is more complex to implement, but may be appropriate in some situations. For example, perhaps the prices in a product catalog are updated everyday, but the product catalog is very large and it takes a long time to synchronize all the records. Field level tracking would enable just changes in the price field to be given to recipients instead of the entire product record.

Change logging. This tracks every change that is made to the data at the field level, including the history of changes. This is useful if the timing of changes is important. For example, the status of an order is updated many times from being created to submitted to picked to shipped to received. It may be important for this change history to be given to the devices so that sales personnel can recount the order lifecycle to their customers.

Three types of changes are detected and given to recipients: create, update, and delete. A challenge with any incremental synchronization (one-way, two-way, or multi-point) is detecting deletions in the original data. Tracking changes that are the creation or updating of data is easy: Put a flag on the data. If data is deleted, there is no data to mark, and therefore nothing to indicate that deletes should be cascaded out to recipients. Imagine that a product was deleted from the catalog. That delete is very important to give to sales personnel so that they don't place orders for the product. However, that requires the catalog to keep track of the deletion until the change is distributed to all sales personnel.

In general, the one-way incremental method is appropriate for

- Data that has only one publisher and one or more subscribers, that is, only one data set is changed, and then pushed out to other recipients who cannot themselves change the data.

- Data sets where only a few records change between synchronizations. For smaller data sets where most fields in a record change, use record level tracking. For larger data sets where only a couple of field values change, use field level tracking. For data sets where the history of changes is important, use change logging.

Two-Way Incremental

This synchronization method is similar to one-way incremental, but data sets on both sides of the synchronization can change. Another term for this is *master/replica*: The master is the central database, and the replica is on the device. For example, at MFG Inc., customers may be contacted by sales personnel in the field and customer service personnel in the main office, and both groups use the SFA system to manage their contacts. Changes made by either party must be given to the other during synchronization (see Figure 7.9).

The obvious challenge is when both parties modify the same data, the master and replica, creating synchronization conflicts. When the synchronization occurs, what happens to the two changes? Business

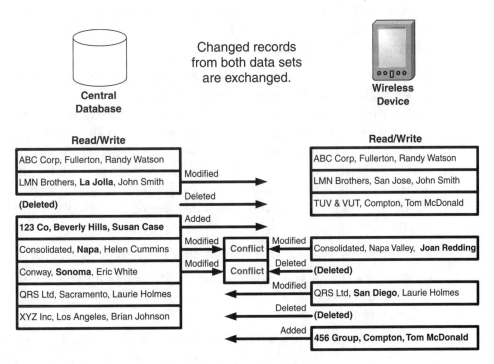

Figure 7.9 Two-way overwrite method of synchronization.

rules must be established for conflict resolution between the master and replica, because they *will* occur.

- Does the latest change win?

- Does the master or replica always win?

- Does one type of change override another, like deletes override changes, or vice versa?

- Do the changes from one organization always override the other organization?

- Is there business logic that can merge the changes from the two sources into one current piece of data?

- Is the conflict flagged for manual intervention?

Another issue with two-way incremental is the perceived conflicts by multiple users. In this situation, many replicas synchronize on their own schedule with the master. For example, two salespeople might change the phone number for a customer and separately synchronize it to the central database at the main office. Each change overwrote the value in the central database and did not create a conflict. However, when one of them checks the value in the central database, they will see that their change has been overwritten, leaving them to wonder what happened to their original change.

To control the issues arising with multiple replicas to a master, enterprises need to implement strict rules about rights to change data and tracking of who does the changes. In our previous example, perhaps only one salesperson should have been allowed to change the phone number for a customer, while the other was granted read-only access to the data. Or perhaps the change should have been stamped with the person's name and date so that each of the sales people would have understood that their change was more recently overwritten and by whom.

In general, the two-way incremental method is appropriate for

- Two data sets against which changes are independently made. Special consideration must be given to conflict detection and resolution.

- Data sets where only a few records change between synchronizations. For smaller data sets where most fields in a record change, use record level tracking. For larger data sets where only a couple of field values change, use field level tracking. For data sets where the history of changes is important, use change logging.

Multi-Point Incremental

Although the two-way incremental method enables two or more replicas to synchronize with the master, only two data sets are synchronized at any one time (see Figure 7.10). This simplifies implementation of synchronization as well as conflict detection and resolution. The middleware need only be concerned with comparing two pieces of data at a time and following simple rules for whose change wins.

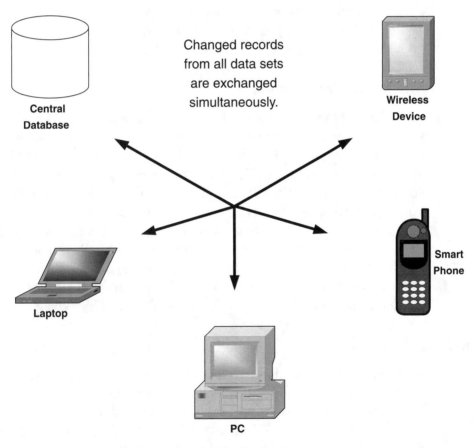

Figure 7.10 One-way overwrite method of synchronization.

The multi-point incremental method of synchronization is used when three or more data sets must be synchronized at the same time. An example is when a salesperson synchronizes his contacts with the SFA application at the main office, Microsoft Outlook on his desktop, and his wireless device *all at the same time*. Special consideration for conflict resolution is critical because the number of data sets involved in the synchronization magnifies the opportunity for conflicts.

In general, the multi-point incremental method is appropriate for

- Changes that are made to two or more data sets that must be exchanged with each other during synchronization. Special consideration must be given to conflict detection and resolution.

- Data sets where only a few records change between synchronizations. For smaller data sets where most fields in a record change, use record level tracking. For larger data sets where only a couple of field values change, use field level tracking. For data sets where the history of changes is important, use change logging.

Synchronization Summary

As described previously, a myriad of considerations exist in designing synchronization. Although I've kept the discussion at an academic level, many wireless middleware products on the market provide actual synchronization services. These packages can offer an out-of-the-box solution to manage synchronization and the extensibility to adapt to your enterprise's maturing needs. Some popular products include ScoutSync by Aether Systems, iMobile Suite by Synchrologic, XTND-Connect products from Extended Systems, and IntelliSync and Satellite Forms Server products by Puma Technology. Synchronization services are also bundled with database systems such as Oracle 9i Lite by Oracle and iAnywhere Solutions' mBusiness Platform by Sybase.

Messaging

Messaging can be used in wireless systems to send email, push out alerts, make remote procedure calls, or even submit transactions. Wireless middleware can offer several features to enhance message handling. These features are similar to those offered by traditional MOM, but are optimized for the low bandwidth and disconnected use associated with wireless systems:

Packaging of many small messages into a single message reduces the number of acknowledgements required and enables efficient use of bandwidth by reducing the number of transport sessions required.

Segmentation and reassembly of a large message into several smaller messages will reduce message size to adjust for harsh communications conditions.

Message filtering enables the application developer as well as the administrator to set parameters, such as message size and source, to control traffic over wireless networks.

Prioritization of messages enables higher priority messages to be processed first.

Delivery/non-delivery notification can be provided to notify both the client and host application when messages have been successfully delivered. It can also return messages that cannot be delivered for any reason.

Retry management will re-send messages if their receipt has not been acknowledged within a predetermined time period. Wireless middleware can adjust this time period on-the-fly to account for harsh communications conditions.

Duplicate detection prevents messages that are resent from appearing as duplicates if the original eventually arrives. Latency times on wireless networks can sometimes be measured in minutes, making this feature crucial.

Message lifecycles can be specified by the wireless middleware so that it will notify a user if his message cannot be delivered within a certain period of time. This can enable the use of business rules for escalation or other intervention.

Messaging services have grown out of the need to improve reliability of communications between applications that are connected. However, in today's wireless environment where applications are more often disconnected than connected, messaging services can be used to enable communication between applications that are disconnected by basically storing messages until a connection is available. Email, instant messages, and text messages are obvious applications that benefit from this simple use of a message service. However, if messaging is combined with a couple of other concepts, its value in wireless systems is magnified.

Transactions

In our MFG Inc. example, we've talked about submitting orders to the central database. If we have a connection to the main office and a session open with the database, the wireless application simply starts to add pieces of the order to the database. For an order, these pieces might include the customer name and ship-to address, payment terms, and of course, all the line items of the order. As each piece of the order is added to the database, the wireless application waits until it is done before adding the next piece.

Let's imagine that we lost our connection in the middle of submitting the pieces of the order (see Figure 7.11).

How should the database react? Should it keep the partial order? Probably not because the original intention was the full order. Should the database roll back the pieces of the order that were submitted and

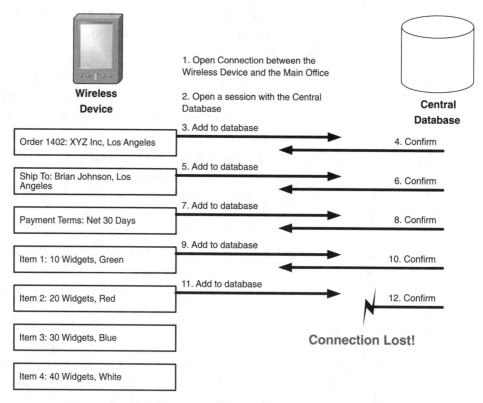

Figure 7.11 Lost connection in the middle of adding an order to the central database.

wait for the wireless application to try again? Probably because the only information it has is a partial order, which we've decided is invalid.

An alternative method for adding the order to the central database is to submit a transaction through a message service (see Figure 7.12).

The wireless application could package all the pieces of the order into a transaction and send it as a message to the database. The message service in turn has the responsibility to open the session with the database and interactively add pieces of the order. The advantage of the *transaction in a message* approach is that less wireless connection time is required, and therefore, the success of the process is less susceptible to a dropped connection. The connection need only survive long enough for the transaction message to be be sent to the message server instead of the entire time it takes to interactively add pieces of the order to the database.

Push and Pull

It is important to think about the two-way flow of messages in a wireless system to fully appreciate the value of a message service. Often this two-way flow is referred to *push and pull*.

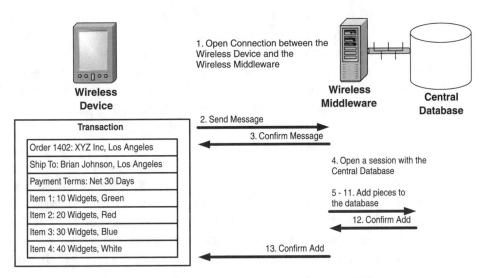

Figure 7.12 Transaction for adding an order to the central database.

Messages can be initiated by a sender to push data to a recipient. Imagine MFG Inc.'s wireless ordering application pushing an order transaction to the central database or an alert message about a product shortage being pushed from the central database out to a vendor's wireless device.

Messages can also be initiated by the recipient to request data, or pull it, from the sender. The wireless ordering application might request statuses of recent orders from the central database, or the central database might request location information from a GPS receiver built into a delivery vehicle's wireless device.

Messaging Summary

Similar to synchronization, messaging services can be very complex in their implementation and operation. The discussion of messaging services was purposefully kept at an academic level. For information about packaged products offering messaging services for wireless systems, check out AIM by Aether Systems, XTNDConnect from Extended Systems, ViaXML and CEFusion from Odyssey Software, AirBoss from Geoworks, Axio platform from Broadbeam, and IdenticonDB from Thin Air Apps.

Management Tools

Management tools are used by enterprises today for central administration of their distributed computing resources. The functions that management tools provide are equally valuable for wireless systems. In fact, management tools for wireless systems are already being integrated into traditional enterprise solutions like Microsoft Management Console, Computer Associates' Unicenter, and HP Openview.

Security

Security of a system can be enhanced in three ways with wireless middleware. First, as mentioned in the communication service previously, data encryption can be used on data that is being sent and received over a connection. Several industry-standard encryption methods can be used such as ECC, Triple DES, Secure Sockets Layer (SSL), and certificate management.

Second, wireless middleware can authenticate users against enterprise control lists before initiating communication. Authentication can be done against

- Existing database user accounts
- Existing network resources like Windows NT user accounts or Lightweight Directory Access Protocol (LDAP) server entries
- Existing application resources like user accounts in ERP, CRM, accounting, intranet systems, or in the wireless middleware application itself

Third, wireless middleware can automatically log and collect statistics from all service-related events in the wireless system. This data could include information on the functions performed on the user's device, the success or failure of transactions initiated, right down to the bytes transmitted between the wireless device and back-end application. This information can be used to track security access, monitor the effectiveness of the wireless application and wireless middleware services, or even generate customer-billing records.

Backup and Restore

The value of backup and restore services will become crystal clear the first time you forget to recharge the device, or even worse, lose it. Wireless middleware can automate backup services to provide the safety your data needs and provide your enterprise the ability to recover from life's little disasters.

Electronic Software Distribution

Wireless middleware can automate software distribution to make sure wireless devices have the appropriate, up-to-date applications installed. More than just an installation tool, this service can be used to configure new devices by distributing several software packages at initial connection, as well as to remove outdated or rogue applications when they are detected.

Wireless middleware accomplishes this in one of two ways. If a new version of an application is available, one type of middleware will simply push out a copy of the entire application. The other type of middleware uses a technique called *byte-level differencing*. In byte-level differencing, each piece (or byte) of the application is compared. Only if the bytes differ does the byte get sent to the device. This cuts down on total data being transferred wirelessly.

Asset Management

Asset management will enable you to report on the types and number of devices your organization uses. Wireless middleware can capture information during a connection that aids in asset management. A collection of data, such as device model, memory and storage capacities, Registry content, user details, and installed applications, is gathered and placed in the middleware server's database for later use.

Scalability

Although not a service in and of itself, scalability enables deployments of wireless middleware to grow as wireless system needs grow from dozens up to thousands of users. One wireless middleware server cannot support thousands of users. Therefore, the ability to cluster a group of wireless middleware servers becomes crucial for medium and large enterprises.

However, a single deployment of wireless middleware servers could be leveraged to provide the previous services for more than one wireless application. For example, an enterprise can leverage one wireless middleware architecture for messaging and synchronization across wireless systems for sales force automation, field data collection, fleet dispatch and executive decision support.

When Does Wireless Middleware Operate?

All of the services mentioned previously operate only when the wireless middleware is operating, which is to say only when wireless and back-end applications are communicating. How often does that occur? It seems that the messaging service described previously would only be useful if communications were kept open constantly. Or would it actually provide some value if communication happened in infrequent bursts, such as one per day or even once per week? Other previous services, such as synchronization and backup, seem particularly useful only because communication happens infrequently.

A fundamental challenge in designing a wireless system is to determine how often communication should occur. The answer isn't always the more, the better. For example, my clients almost always assume that they need constant communication between their wireless device and their back-end application. In reality, they most often need a store and forward architecture because their data is not as time-sensitive as they think. Furthermore, constant communication is cost-prohibitive and, at least with wireless WAN technology, nearly impossible to achieve given the slow bandwidth, limited coverage, and dropped connections associated with today's technology.

Let's discuss three options for how often communications could occur in a wireless system: in real time, store while disconnected and forward when connected, and the happy medium I've dubbed *hybrid*, which leverages the advantages of both real time and store and forward.

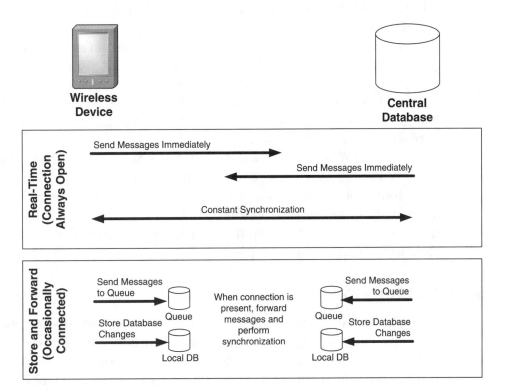

Figure 7.13 When wireless middleware communicates.

Real Time

In a real time model, wireless and back-end applications are constantly connected while operating. A wireless device operating in this mode is similar to a computer sitting on a corporate LAN. When you access a database from a computer on a wired network, your request is handled immediately. Data in the wireless real time scenario is accessed directly at its back-end source and is always up-to-date, thereby eliminating the need to synchronize data to the device. A thin client (such as a Web browser) can be deployed on the wireless device that leverages the business logic of the back-end application. Transactions and messages are sent and received immediately.

However, wireless connections today contain various degrees of limitations and instability. Wireless applications that utilize a real time model should only be implemented in environments that can provide appropriate connection reliability. Appropriate doesn't mean best; it just means that the reliability needs to be sufficient for the wireless system. Wireless WAN's offer the greatest degree of location flexibility, but are fraught with the most severe challenges: dropped connections, low bandwidth, and dead spots in coverage. For some applications, it may be sufficient to get occasional connections for short periods of time. Or the choice could be made to sacrifice mobility for greater reliability and use wireless LAN's, but they still have the same challenges, just to lesser degrees. Connections are rarely dropped unless the user wanders into a dead spot in coverage, and bandwidth is significantly higher than wireless WAN's. Using wireless piconets is simply taking the last step back to the user's desk and substituting radio frequency for a three-foot cable. Finally, connecting a wireless device to a wired modem, Ethernet cable, or cradle will still technically give you real time communication with ultimate reliability, but at the cost of complete immobility.

Store and Forward

Store and forward wireless systems are a return to client/server computing with a new caching layer in the middle. The client and server do not have to ever be connected while operating. This model uses a thick client that has embedded business logic and stores data locally so that it can operate in a completely disconnected mode. Updates made on both the client and server sides are stored until a connection is made. Although the connection is open, communication is purely to forward the stored updates. The forwarding of these stored updates could be just the emptying of message queues that have built up or perhaps full-scale data, file, and Web content synchronization. Once all changes have been forwarded from the client to the server and vice versa, the communication is complete and the connection can be broken.

The challenge with the store and forward model is that the client and server must operate for some period of time without knowledge of one another. Messages and transactions are not delivered, and neither has access to the most recent data or applications on the other. This situation may be sufficient for certain wireless systems. For example, consider an application on a wireless device used to capture monthly utility meter readings. Is it reasonable to expect that a user could perform several meter reads, and then forward the results at the end of the day, or even at the end of the week? This question can only be answered by the enterprise implementing the wireless system.

Hybrid

A more thorough, yet complicated model is one that is a hybrid of real time and store and forward communication. A hybrid model has portions that communicate in real time as needed, and other portions where updates are stored until they can be forwarded. For example, consider an application that assisted with the repair of utility meters at their location. Upon initial disposition of the meter's problems, information could be captured about the necessary tools, parts, and resources necessary for repair. This information, in the form of a maintenance or issue ticket, is communicated in real time to a back-end dispatching application. In order to complete the maintenance ticket, the user may want information about the particular model of meter, so they look up the latest reference materials that are stored or cached locally on the device. This lookup information was forwarded to the device during the last synchronization. Since it's relatively static and present in a large quantity, it is stored on the device for quick access.

Where Is Wireless Middleware Deployed?

Following the lengthy discussion of how wireless middleware manages communication and what services it provides, it is natural to assume that wireless middleware is a richly featured, highly complex, prohibitively expensive piece of software that drains tremendous computing power from only the beefiest enterprise servers. To implement everything we've discussed so far, that would be true.

However, in practice, wireless middleware can provide any or all of the features and services discussed. (Of course if it provided none of them, it wouldn't be middleware, now would it?) To understand the variation in wireless middleware implementations, let's take a look at where they can be deployed.

Wireless Device

Wireless middleware can be deployed directly on a wireless device. Given the computing power of these platforms, it is easy to understand that these could be some of the smallest implementations. Examples of middleware on the wireless device could include

- A connection monitor that switches from a more costly and unreliable wireless WAN connection to a wireless LAN connection when it becomes available

- A message queue that sends HTTP requests to a back-end application, or stores them until a connection is available

- A synchronization engine that, upon user request, seeks out updates in back-end databases and requests that update transactions be sent to the wireless application

Deploying wireless middleware to the device itself can introduce new problems: maintenance and support (see Figure 7.14). The wireless middleware is deployed to every device introducing dozens, hundreds, or thousands of potential points of failure, not to mention the same number of wireless middleware administrators. However, if users only have wireless devices and no PC or server resources, wireless middleware on the devices may be the only alternative.

PC

You are already familiar with wireless middleware running on a personal computer if you have a Palm OS or PocketPC device hooked to it. Palm's HotSync Manager and Microsoft's ActiveSync are both examples of wireless middleware. These programs manage the communication

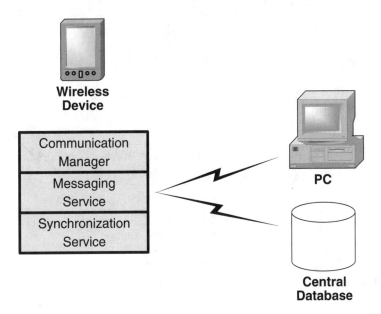

Figure 7.14 Wireless middleware on a wireless device.

between device and PC over infrared, serial cable, or USB cable. (To do this, a small portion of the wireless middleware is also deployed to the device itself to manage the communication from the device side.) These packages provide a variety of services that we described previously, including

- Messaging (send outbound and receive inbound email)
- Synchronization (address book, notes/memos, calendar, and tasks)
- Backup (databases, applications, and configuration)

With both HotSync Manager and ActiveSync, developers can write plug-ins that provide new services such as distributing a department's sales data to the device or caching HR Web pages from the corporate intranet for offline viewing.

Wireless middleware in the previous examples isn't really overcoming challenges of wireless connections; it is just serving the needs of the wireless device and applications. In the case of maintenance and support, the PCs are more familiar to enterprise IT departments than are wireless devices (see Figure 7.15).

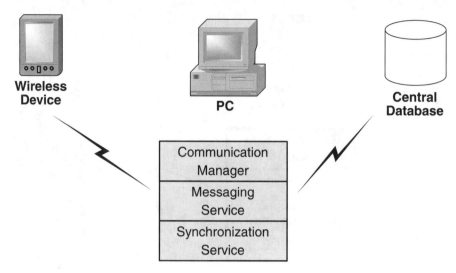

Figure 7.15 Wireless middleware on a PC.

Server

Most wireless middleware in the enterprise is deployed on a centralized server. This facilitates its management by IT resources, reduces the distribution of middleware updates, and leverages more powerful computing resources. In fact, almost all commercially available wireless middleware packages for the enterprise are designed to run on a server. (Several commercially available wireless middleware packages for consumers are designed to run on a PC.) As with PC wireless middleware, a small portion of the wireless middleware is also deployed to the device itself to manage the communication from the device side.

Most commonly, the wireless middleware is addressable as an IP address and port, so making connections from wireless and back-end applications is very straightforward. Any medium that can establish an IP connection will work: dial-in or wireless modem, wired or wireless LAN card, or even a proxy through a PC. Some wireless middleware does not use an IP connection. Instead it addresses and routes data using its own protocols that cover layers three and four in the OSI. Examples of wireless middleware using non-IP-based connections include WAP and SMS gateways.

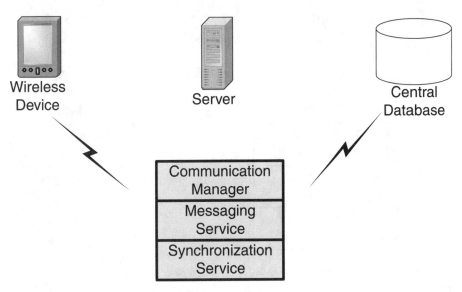

Figure 7.16 Wireless middleware on a server.

Peer-to-Peer

Most wireless middleware solutions today extend a client/server paradigm: Full data resides on a server, and a subset of the server's data is distributed to a client. In the enterprise, the wireless middleware frequently resides on a server with only a small communications manager on the device. However, there is a growing demand for peer-to-peer capability changes that affect the client/server paradigm slightly.

The main concept behind a peer-to-peer model is that communication is enabled between devices without need to route it through a central server for retransmission to the recipient (see Figure 7.17). In effect, each device is both a client and a server. A request from the client on device A is routed to the server on device B, which responds directly back to the client on device A. The uses for peer-to-peer are now becoming prevalent in the PC world where one device will leverage the computing power, storage, applications, and data of another. Napster brought the peer-to-peer model into the limelight in 1999. New applications for peer-to-peer that automate traditional workflows in ways that improve worker productivity will continue to drive its use.

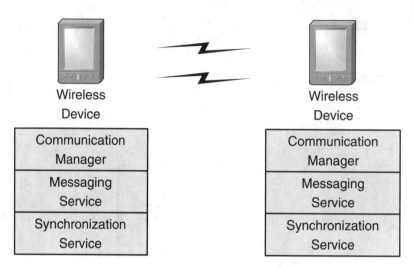

Figure 7.17 Wireless middleware for peer-to-peer communication.

Considerations in Implementing Wireless Middleware

Other sections of this book cover issues about strategy, development, deployment, and maintenance of wireless systems. However, I've captured here some of the common considerations specifically regarding wireless middleware.

Buy Versus Build

As with any enterprise IT project, you should seek to buy a solution before jumping in and building it. When you build a solution, you become a software development company with a customer base of one, which takes a very large business need to justify.

Several wireless middleware vendors have products on the market today, and more are entering the space every month. Here are some qualities to look for in your wireless middleware vendor:

Platform support. Do they support your selected platform and maybe even others you may consider in the future? Some support all four platforms and others support one.

Network support. Same question about network as for platforms. Consider wireless WAN, LAN, and piconet technology that you have or will have implemented for your wireless system.

Specific versus generic services. Are you looking for a product to synchronize Lotus Notes data from a PC to a Palm Vx via a modem and a CDPD network? Or are you looking for a broad synchronization engine for a variety of enterprise data? Your answer to this question will point you in the direction of certain vendors and steer you away from others. Before you default to generic to give yourself flexibility for future services, remember that generic solutions will require more development and maintenance effort to deliver the one or more specific services your enterprise needs.

Scalability and reliability. Accurately assess the current *and* future needs for performance, and then make sure the vendor can deliver. Examine your growth path and timing carefully. It may be more cost effective to purchase a lower priced, specific solution today, verify the expected benefits of the wireless system during a pilot period, and then switch to a more robust wireless middleware solution later.

Technical support. Evaluate technical support programs and quality when choosing a vendor. In addition to investigating technical support policies, such as hours, response times, and escalation procedures, you should also look at what kind of support it offers to developers through a developer program.

Value-added services. If this is your first wireless system, you're going to need help. Some vendors of wireless middleware also sell other enterprise products and even professional services like training, system design, application development, systems integration, deployment assistance, or application hosting. Outside consulting and development groups must have experience deploying multiple types of synchronization servers. They must also understand the challenges of synchronization.

Price. This is the big question. Start with an internal assessment of the return expected on your investment. Have you quantified the benefits you will reap with the wireless system? How much of the wireless system cost will be the acquisition and maintenance of the wireless middleware? Then work with the vendor to understand their pricing model. Does it fit well with your business? How willing is the vendor to work with you? Does the price include per processor installation, seat licenses, concurrent user licenses, named licenses, support and upgrades, and so on?

Resources for Wireless Middleware

The team for developing wireless middleware is extremely different than the one for acquiring it. To develop, your team must be proficient in all areas covered in the previous buy versus build discussion. Remember, building a solution internally is like becoming your own software development company. This software development company should have detailed knowledge and experience with all layers of the OSI for the platforms and networks you will be deploying. They should have experience developing services you will need like security, messaging, synchronization, and so on. Finally, they should have programming experience on the targeted platforms, not just for the OSI, but for user interface, business logic, and data management. Sound like a lot of experience to grow internally? It is, but if no products are on the market that meet your needs, you may develop the solution internally.

The team for implementation and support is probably closer to the skills you already have on staff: familiarity with your networking environment, device platform(s), and back-end platform(s). Any acquired wireless middleware will come with an implementation plan, either as written instructions for simpler products, or as professional services baked into the price for more complex products. If the product is so difficult to use that after its implementation your folks still don't have a clue how to support it, then the vendor has done a poor job transferring knowledge to your staff. Unfortunately, with the relative immaturity of the wireless space, you will find products that may be technically amazing, but unbelievably frustrating to set up, use, and support.

The Future of Wireless Middleware

Several technological and sociological changes will occur in the next year; some will be evolutionary and others revolutionary. Here are a few that I predict will have the most profound impact on wireless middleware.

Mobile IP

In the next year, devices will have the ability to communicate over more than one wireless connection. For example, a device in a vehicle may communicate over a wireless WAN when out on the road, but over a wireless LAN when the vehicle is at headquarters. Another example is a handheld device communicating with the PC via a wireless piconet, but switching to the wireless LAN when picked up and carried away. Although it is technically possible to do this today, it is costly to establish one wireless connection, let alone two or more.

One thing today's IP networks don't do well is adjust moving devices to different subnets or entirely different networks. A developing standard called Mobile IP created by the Internet Engineering Task Force (ITF) will enable devices to dynamically move without powering down, closing a session, or dropping a connection. The impact of this development is the further transparency of wireless middleware to users. They will neither concern themselves with the differences between wireless WAN, LAN, and piconet communication, nor with the switching between them.

SyncML

Currently, a variety of synchronization protocols are implemented in both wired and wireless middleware. Although these multiple forms of synchronization are currently being driven by varied and competing forces in the computing world, it will be difficult for true ubiquitous computing to proliferate if there isn't a unified standard for synchronization.

SyncML (www.syncml.org) is an industry initiative headed by IBM, Lotus, Motorola, Nokia, Palm Inc., Psion, and Starfish Software with the purpose of developing a singular industrywide synchronization protocol. SyncML in version 1.0 can manage the device and network independent synchronization of any personal information including email, address books, calendars, and task lists. Future versions will support generic synchronization of any enterprise data.

The SyncML initiative has specified two protocols. The synchronization protocol defines how a SyncML session is initialized and how information used for connecting to the remote server is represented. The representation protocol, which is an XML markup language, defines the format that must be used to represent item data during synchronization.

Clearly, an opportunity for SyncML to shape synchronization services in the future of wireless middleware exists. Imagine if setting up synchronization was as simple as naming the source of data and your device simply synchronized: no concerns about mapping data sets, replicating data, or configuring conflict resolution. The end result is that synchronization may eventually be transparent to enterprise IT users, developers, and administrators.

The Bridge from Here to There

Currently, about 20 major vendors of wireless middleware with literally hundreds of minor players are in the space. The products from major vendors offer a comprehensive treatment of one (or even two) of the services described previously, such as complete suites of synchronization services and management tools for PocketPC, Palm OS, and RIM devices. Although feature-rich, these products tend to be very generic and require lots of customization before they add value in an enterprise.

However, most vendors offer a very specific treatment of one service, such as synchronization of contacts, to dos, and calendars between Palm OS devices and Lotus Notes databases. This is not to say that narrowly focused products are not high quality; they just do a few things and do them well.

By the end of 2001, I see the number of vendors and products in the wireless middleware market peaking. At that point, wireless middleware will be beyond its early adopter phase for enterprises and into mass adoption, driven by the combination of an economic recovery (we all hope) and the availability of 2.5G wireless WAN technologies. As more enterprises begin to consider wireless middleware, vendors that are unable to provide products that are feature-rich, scalable, and offer out-of-the-box value will be unable to compete. A few that have great underlying technology for specific functions may be snapped up by the big guys, such as support for a device or database that can be melded into a larger product.

Beyond 2001, the market for wireless middleware will continue to mature:

- Synchronization and messaging will continue to mature to maximize throughput and reliability of wireless networks.

- Management tools will become more important as wireless devices become prevalent, even pervasive, within enterprises.

- Emphasis will be on real time and hybrid applications, with little value placed on those products that handle store and forward architectures only.

Wireless middleware or the magical rainbow, whichever you prefer, is a complex albeit crucial piece to an enterprise wireless solution. With the conclusion of this chapter, we have covered all of the pieces of the wireless architecture. The next stop is handheld application design. The tools used to create cutting-edge wireless applications on the handheld aren't exactly ones you or your organization are likely familiar with. So, we'll look at the toolsets available to you and some general tips for creating handheld applications.

CHAPTER 8

Handheld Application Design

A ny enterprise IT manager knows that computing technology is a tool for delivering business objectives. You want to spend the majority of your time solving business problems and not fighting technology. In developing business applications, you desire tools that aid in the delivery of the solutions in the most rapid manner while delivering the functionality and stability your users expect. Microsoft developed Visual Basic in part to speed the development of business solutions. Spend less time on the getting the technology working and more on implementing your business logic.

The same approach should be taken in the wireless arena. Some of the first tools delivered by device manufacturers were unstable and lacking in features. Today the tools are much more robust and plentiful. These technologies range from simple forms packages for collecting data to full-blown development environments. This chapter covers your choices for designing and developing handheld and wireless applications so your organization can spend more time solving business problems and less time fixing technology issues.

Whichever tool or set of tools you determine are optimal for your organization, you must keep in mind some basic design considerations for your applications. This chapter also helps set the parameters for the types of applications and functionality you can deliver with today's packages.

The Great App Debate

A great debate is underway for wireless and handheld application development. The debate isn't so much out in the open at conferences or on message boards; rather, it's implied by the variety of technology wares brought to market and the companies that are purchasing them. Some claim to be agnostic on this issue. Others say they support both. In the middle exists a large contingent that has set course with their strategy of one or other. This issue applies to both the technology companies building these applications and tools and the enterprise that must choose a course of action. That debate: whether to build handheld and wireless applications leveraging Web technology or the more traditional computer languages like C, Visual Basic, and Java.

This distinction and topic is critical to any enterprise's deployment. Some great ideas have gone bad because of how they were implemented. We're going to spend time in this chapter to explore the development styles that are available for handheld and wireless computing. The explanations will be in layman's terms with technology terms thrown in when necessary. It will give you a solid foundation on how your applications should be built. In fact, it is never as easy as saying, "If you are creating a sales force automation tool, use Web technology," or "If you need a patient tracking tool, use a traditional computer language." The decisions differ based on the specific company, the vertical market they are in, the extensiveness of the functionality they need, the devices being utilized, network coverage, and a host of other items.

Native Versus Web Architecture

Before we can begin our pro/con discussion, we need to define a couple of terms. The first is *native application*. There are likely 100 other terms for native applications. I've chosen native applications, but you could have chosen another like compiled or executable. A native application in the context of this book is an application written for a particular operating system. It is programmed in a traditional computing language like C, C++, Java, or a Rapid Application Development (RAD) tool (such as Visual Basic). Development of solutions requires the typical skills you would find in a good developer. The end result of creating a native application for a PC is typically a file with .EXE on the end of it. For example, when you run the game Freecell (see Figure 8.1), which ships with Windows, you are running the native application, Freecell.EXE. Native applications can also be built for Palm, RIM, and PocketPC using a variety of software development packages.

Contrast this with solutions developed exclusively with Web technology. These types of applications will use things like HTML, WAP, Web servers, Active Server pages, Java, and supporting graphic design tools. In a sense, it's using the skills and technology that present data and graphics within a Web browser either from the Internet or your corporate intranet. Figure 8.2 shows the login screen of Yahoo's Web-based e-mail application. The browser displays the interface while the logic and data of the application reside at Yahoo!

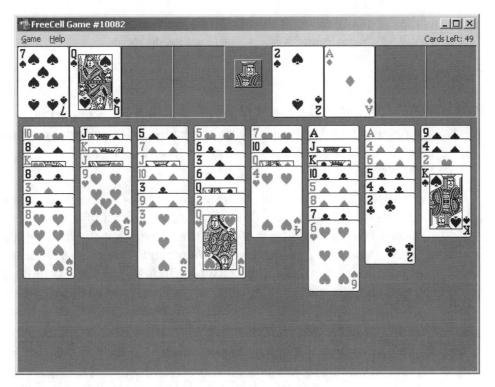

Figure 8.1   Freecell is an example of a native application.

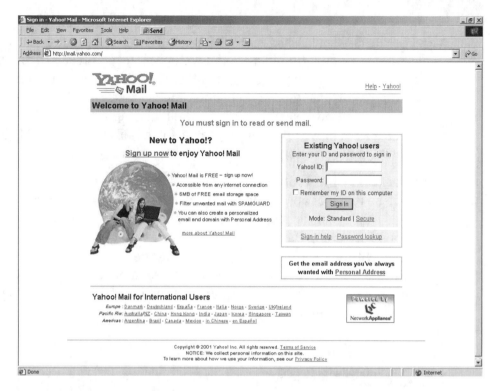

Figure 8.2 Yahoo! e-mail is an example of a Web application.

A number of fundamental differences exist between these two approaches (see Figure 8.3). Think of handheld-based applications in three distinct parts independent of whether it is a native application or a Web application:

User interface. This is the view the enterprise worker or customer (if an external facing) has of the application. It is all of the buttons, lists, tables, tabs, text, and graphics on the screen that the user interacts with. Some user interfaces on certain handheld devices are more robust and feature-rich than others. Windows CE, for example, utilizes three-dimensional buttons, tabs for organizing data, and an extensive menu system. Palm has a more basic interface that supports buttons and checkboxes, but few advanced controls.

Business logic. Business logic is the code that performs all of the data calculations and information processing, decides which step of the business process to handle next, and how to format data on the

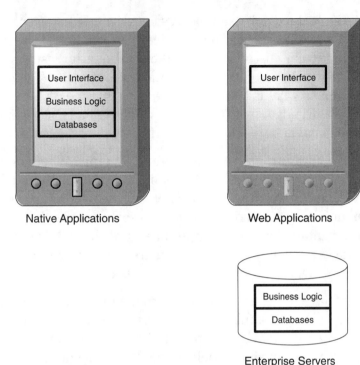

Figure 8.3 Overview of application development approaches.

screen. When a developer programs an application, a great chunk of his or her time is spent in this area.

Databases. This is the location from which data is read, written, updated, deleted, and created. The business logic is responsible for deciding which of these operations to perform on the data. Databases can be stored on the device for speedy retrieval, on a corporate server, or some other remote location.

Native applications typically have three of these components on a device for maximum performance. During synchronization, business logic and data on the back end may update the business logic and data on the device. However, when the application is in use, the source of data and logic will be on the device itself. Web applications are in most cases, again, opposite to this. The business logic and data are somewhere else, which is typically why a true wireless connection is needed for the application.

Let's look at a quick example. A sales account manager is visiting a customer site with his handheld device. He would like to review the sales of this particular customer during the same quarter last year. The account manager makes a query through the user interface. The *business logic* determines what is being requested and that the request is valid. It then retrieves the data from the database, determines how it should be formatted, and updates the user interface with the sales history.

The overriding theme of this section is simply: Where do we place each of these three blocks? Native applications store all three pieces on the device itself. Web applications present the user interface while the remaining logic and data reside on a server somewhere else. Each of these handheld applications architectures has advantages and disadvantages. Depending on what you are attempting to accomplish, you'll choose one, the other, or in some cases both.

Now I have to insert a disclaimer to the previous statements so I don't take a literary beating from my peers. The previous model is a simplification of handheld application design, but it is fairly accurate. For the most part, the business logic in the Web applications model resides on a Web server somewhere else. Each interaction with the user interface forces the device to send a request to this Web server. The Web server runs various scripts or programs (the business logic layer) to access databases and return results.

Technologies do exist for adding business logic and database storage to a handheld device in the Web application model. JavaScript enables for business logic within a Web page, but only a small minority of devices has the ability to handle JavaScript due to its processing and memory requirements. AvantGo Server by AvantGo is one company leveraging Web technology and applications in their product. However, they've built products that enable for fairly robust business logic and database capabilities on the device itself.

Additionally, not all native application solutions will have their entire set of business logic on the device. Handheld solutions frequently require data from a number of data sources. This immediately implies other databases and business logic located somewhere else. Native applications may use database and business logic stored locally and retrieve other parts from a corporate network via a wireless connection. All of these disclaimers and caveats aside, the model holds in 90 percent of all cases.

Native Versus Web: Pros and Cons

We'll now explore some of the decision points for choosing one model over the other. Keep in mind that a complete enterprise wireless strategy can include multiple business systems and applications. It may be that some are better served with native applications and others are better supported with Web applications.

Issue 1: User Interface Expectations

When designing for handheld devices, the user interface becomes as important as any business process you are trying to move to wireless. Screen real estate is extremely limited and therefore, clever designs must be used. The operating systems of the handheld devices we're discussing have fairly robust user interface elements available to them (see Figure 8.4). Controls include those you are familiar with on a desktop computer (buttons, graphics, listboxes, and so on) as well as others unique to the handheld device itself: vibrating alarms, hardware buttons, and jog dials.

Web applications are much more limited at this time. User interface elements are limited to those supported by HTML and WAP. They include basic elements: buttons, edit boxes, checkboxes, and so on. While all types of data you would want to collect can be handled by standard

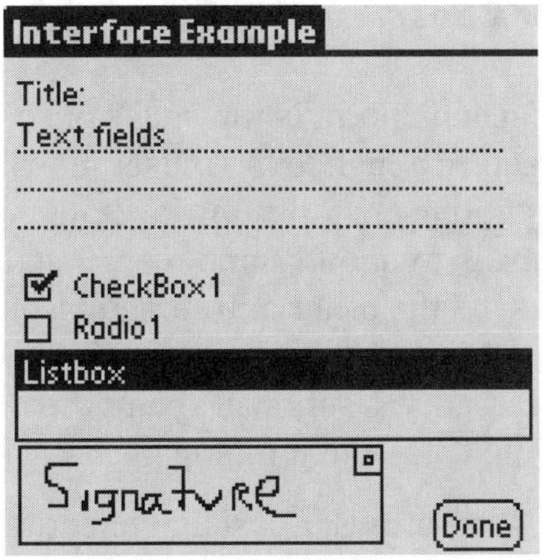

Figure 8.4 Examples of user interface elements available in a native application.

user interface elements, you'll find them extremely limiting as the optimal method in many cases.

A home healthcare worker frequently owns the task of recording the vital signs of the patients she visits. Although an edit field would suffice for collecting the data, it requires that the nurse use the handwriting recognition of the device or type it on a pop-up or hardware keyboard. This may be daunting to someone who is not comfortable with computers. In an application-leveraging Web technology, you would be limited to these types of fields. Alternatively, a native application could use a slider control on the screen. Think of a slider control as a sideways thermometer. The user can graphically drag the temperature higher or lower, bypassing a more complicated interaction (handwriting recognition) with the device. The slider is intuitive to use and the quickest way to get body temperature or heart rate into the device. Having access to these creative input methods frequently requires a native application.

Please don't walk away from this first issue and believe it is trivial. When designing for small form-factor devices, interface design and creative methods for performing input and output, as well as giving users visual feedback as to where they are within the application, is *crucial* to success. I cannot emphasize this enough. I've seen some great ideas for enterprise applications that were ultimately hampered by the interface.

Summary: If a creative and simple interface is required, a native application gives the most flexibility.

Issue 2: User's Business Process

The second decision point is the user's business process. Does the application require a high level of user interface interaction? Applications like this find users inputting large quantities of data, tapping on interface controls frequently, and resorting data to view from different vantage points. At the other extreme, the solution may be one whose primary method of use is data viewing. Executive information systems fall into this category. Users will spend the majority of time reading reports, viewing data, and jumping between screens.

As with a PC, Web-browser-based applications do not act as robustly as standard applications. Even with a high speed Internet connection, a Web-based application can perform poorly due to the fact that every screen change or update requires a request to and response from a Web server. In the same way, we have found Web-based technology to be inadequate in situations where large quantities of data need to be displayed and heavy input and manipulation of that data is needed.

One of our clients needed a wireless ordering and inventory engine developed to extend their Web site's B2B engine. One of the requirements mandated that hundreds of products be available onscreen for ordering. Due to the amount of data that needed to be displayed, the need to record a large number of inventory and order quantities and the ability to quickly resort the product by supplier, location, or category, the Web application solution they requested did not perform to expectations. We rewrote this order and inventory application as a native application and found the performance dramatically better and the adoption by users more in line with their original expectation.

Summary: Applications requiring high levels of user interactivity perform more robustly as a native application.

Issue 3: Connectivity

What type of connectivity will your users have with their devices? Will it be true wireless over a CDMA network or will it use a store and forward synchronization model? Most Web-based solutions require a true wireless connection to function. What would happen if the worker left WWAN coverage by walking in a building? Each request for information in a Web-based application requires that the browser connect to a Web server. Without that access, the application would error out and the user could lose data he collected previously.

A number of companies who create Web-based middleware for wireless computing have built-in features to accommodate out-of-coverage situations. They typically require some programming to store data on the device when you lose a wireless connection. Also, there will be larger memory requirements to cache all of the pages that you might need in the future on the device. If your company chooses an HTML (Web) or WAP-based solution, make sure the solution accounts for the possibility of being out of coverage.

However, if your company utilizes WLAN (802.11b or a) or Bluetooth connectivity indoors, a Web-based solution works well. The handheld simply becomes the user interface to the business logic and databases stored on or near the Web servers. The performance of these wireless networks is fast enough and coverage is good enough that you can rely on having connectivity as if you were hard-wired to the corporate network.

A help desk application may be one where a Web application would be adequate. Technicians walk through building fixing users' computers. Having access to issue information while in the field would save a huge amount of time. If a company invests in a corporate wireless LAN, a Web-based application could prove to be a quick and robust solution.

Summary: Web applications in most cases demand wireless network access at all times. Native applications perform better for sometimes-in-coverage situations. Web applications should be designed to handle the potential for out-of-coverage conditions as well.

Issue 4: Time to Deliver

All IT managers want to deliver robust, inexpensive, scalable, secure, and reliable applications tomorrow. The reality is that different technologies require different amounts of resources. If a company already leverages an extensive amount of Web technology, it *may* be less costly in terms of time to implement a Web application solution than a native application.

Web-based solutions leverage technologies that are, in general, easier to implement. A corporate intranet may be designed in such a way that inserting transformation middleware to transform Web pages formatted for a 17-inch monitor in such a way that it will look reasonable or even good on a handheld device's screen. If not, designing HTML or WAP pages from scratch may also be faster than creating a native application.

If during development you realize that users will be out of coverage at different points, but still need access to this data, it is then that you must consider Web-based tools that allow you to cache data, business logic, and Web pages on the device. When you have to design for coverage conditions like this, the gap between building a native application and a Web application narrow significantly.

Summary: Web applications, generally, can be developed more quickly than native applications.

Issue 5: Cost to Deliver

Cost is always a tricky subject with any technology. So many factors go into cost. For different projects, I'll know to the dollar what strategy, design, development, and deployment will cost and some piece of infrastructure or unknown at a client site will throw that estimate off significantly.

With that disclaimer in place, Web applications are typically less expensive to deliver than native applications. If your company already owns and operates a substantial amount of Web infrastructure and back end business logic, it may be possible to leverage that very quickly through a Web application running on a handheld device.

I can't emphasize enough that the design of the interface, the ease of use of the application on the handheld, and the need for wireless connectivity are probably the largest reasons why users will accept the solution or not. Do not take the easy way out here. Do not think that your task is to simply to get the data on the device and you are done. The presentation of that data, ease of manipulation, synchronization challenges, and other factors play a huge role in designing wireless and handheld solutions. Because of this, you need to carefully weigh the pros and cons of Web solutions today.

Summary: With a number of caveats, wireless Web applications can be less costly to deploy than native applications if you can leverage existing business logic and processes.

Web Versus Native: The Winner Is . . .

Based on the five criteria I presented previously, it would seem that I've taken a negative view of Web-based technology for application design and development. On devices with high-speed wireless coverage and powerful processors, a Web application can work well. Unfortunately, in most scenarios, we do not have this luxury. As wireless networks cover more land and increase in speed and device manufacturers design more powerful handhelds, the reality of leveraging Web technology will come become the right choice for many solutions.

Whichever technology you use for application development, keep in mind that you must develop for both out-of-coverage and in-coverage scenarios. If you ask what our company has developed for our clients based on their strategy and requirements, 9 times out of 10 the solution is a native application. As wireless coverage increases in reliability and speed, we've seen more time where Web-based applications can work. Web applications are gaining ground.

A Bit about iMode

iMode is a technology brand name of NTT DoCoMo out of Japan. Its popularity in the American media makes it worth mentioning. Again, the marketing hype and media engines our country thrives on would make you believe that iMode will single-handedly drive adoption of mobile Internet users and come to the rescue of WAP. So, let's take a look at this technology and determine what America might do with it.

DoCoMo is a subsidiary of NTT, the Japanese communication super giant. It runs the largest, most successful cellular phone network. In fact, the number of cellular phone subscribers in Japan is greater than the number of traditional land line users. There are a number of reasons for this, but sufficed to say, the Japanese love their cellular phones. iMode is simply wireless access to custom Internet sites that support iMode phones. Said another way, a user can subscribe to cellular service and then add the iMode feature to have access to the *wireless Internet*.

The iMode service launched on February 22, 1999. By March of 2001, the user base exceeded 20 million users; that's almost 1 million new users per month. Talk about success! DoCoMo reports that over 800 companies provide iMode services and over 40,000 iMode Web sites exist. Compare this to the United States, where wireless data subscription rates languish in the hundreds of thousands or very low millions.

What Is iMode?

iMode is a packet-based Internet technology not too dissimilar from WAP. As you might recall from Chapter 6, packet-based networks possess an always-on feature. Data can be sent and received almost instantaneously. No need to wait for the device to establish a connection like a modem requires. Because iMode is more like WAP or Web technology than a native application, it has all the pros and cons associated with it. iMode relies on a markup language called CHTML, or Compact HTML, an invention of DoCoMo. It is based on HTML and uses a subset of it along with some additional proprietary tags.

Because of the fact that the Japanese already have a 2.5G infrastructure in place as well as cellular phones with large, color displays, they can show more content. Some of the services used in Japan include email, news, sports, as well as services to help people find new friends and download daily cartoons. Having spent some time in Japan and being able to see the Japanese use the service, I'm impressed. Popular with the young to middle-age crowd there, you'll find people using it on trains and while walking down the street. It's as common as seeing someone in the United States take out his cell phone to have a conversation.

DoCoMo recently took a 16-percent stake in AT&T Wireless in the Unites States and is expanding into other coverage hoping to recreate its success there in other countries. Japan's success with iMode begs the question: Why is it so successful in Japan when carriers can do little to get Americans to adopt their services? Some ideas include

Cell phone penetration. More than half of Japan's population uses cellular phones, making them commonplace.

Cell phone cost. The cost of cellular phones is extremely low, similar to the United States.

Cell phone features. Cellular phones in the United States have some interesting and clever features, but pale in comparison to Japan's ultrasleek phones with larger, color displays.

Infrastructure. Japan already has nationwide 2.5G service. Therefore, carriers are able to offer more advanced services.

Microbilling. Billing for iMode, which is a combination of a flat fee and per packet costs to transfer data, appear on the user's phone bill. Similarly, those partner companies offering iMode sites and services for payment can also bill through this system. In other words, the cost barrier to entry for iMode is low.

Fashion. Don't underestimate the power of fashion. On a particular Saturday in Japan, one of the large cellular phone carriers was having an outdoor marketing program complete with models and people dressing as the company's mascot. They were introducing a new cellular phone with a built-in camera and color display. People flocked to get a closer look. It's fashionable to own iMode phones and subsequently use them to stay in communication with friends through voice, e-mail, instant messages, and other types of communication.

iMode in Other Parts of the World

NTT DoCoMo is diligently trying to extend its blockbuster service to other parts of the world. Its stake in AT&T Wireless is evidence of that. Now that we know that iMode is essentially a Web-based service, the question is will America adopt it? Maybe. Today the Unites States has Web-based services for wireless devices, but few people subscribe or use them. However, the services available by cellular phone today are ill-conceived and trite. Navigation through the services is poor and difficult. Screen sizes are small and text-based. Should the Unites States put in place infrastructure like Japan's along with hi-tech handsets, this may change. Once the playing field is leveled in terms of infrastructure, the question will become who has the best, most innovative, and most convenient services? iMode definitely has a leg up in this area. When and if this happens, iMode could take off here too.

Business impact. Now that we know iMode is a Web-based service, we can assess against things like the wireless Web and WAP. If iMode becomes available, businesses will need to evaluate along with other technologies for fit.

RAD Tools

Rapid Application Development (RAD) is any programming system made to get applications developed and deployed very quickly. RAD tools build native applications. They sacrifice performance of the application for ease of design and development. Visual Basic is one such tool. Numerous enterprises use this tool exclusively for business applications. In the early days of mainstream handheld computing (three to five years ago), the choices for development were much more limited. Typically, enterprises were forced to use C or C++, and the tools for building applications had a large amount of instability. Development and testing cycles were long. This has all changed recently (see Figure 8.5). RAD tools, like Aether Systems' ScoutBuilder, AppForge, Pencel's Kinectivity Studio, and Pumatech's Satellite Forms (all for Palm-based devices), simplify the development of handheld solutions for business.

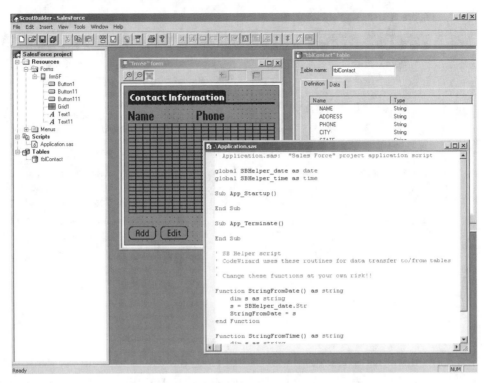

Figure 8.5 ScoutBuilder from Aether Systems is one tool for RAD on Palm.

For Windows CE, Microsoft's Visual Basic fills the RAD gap. RIM is still waiting for such a tool. The wireless Web platform utilizes technologies like HTML and WAP, which already simplify the development process. See Table 8.1 for a list of some of the major RAD players. Palm's RAD tools have evolved greatly in the past two years. Although I won't get into a discussion of which package to choose or whose tools are better suited to your needs, it is worth noting that these development tools fall in the 80/20 Rule.

You can deliver 80 percent of the wireless and handheld applications your business needs with 20 percent of the effort through the use of RAD tools. When you choose to use a RAD tool, you can use a number of implications to your wireless and handheld projects. Let's look at these so you can keep them in mind:

Performance. Performance of applications written with RAD tools is slower than the corresponding application written in C or C++. This is one of the biggest downfalls, if ranked against the others. In many

Table 8.1 RAD Tool Options

TOOL	MANUFACTURER	PLATFORM	URL
Satellite Forms	Pumatech	Palm	www.pumatech.com/Satellite Forms Enterprise.html
ScoutBuilder	Aether Systems	Palm	www.aether systems.com/software
AppForge	AppForge	Palm	www.appforge.com
Visual Basic	Microsoft	Windows CE	www.microsoft.com/mobile/developer/default.asp
Kinectivity Studio	Pencel	Palm	www.pencel.com/products/studio.htm
MobileBuilder	PenRight!	Palm/Windows CE	www.penright.com

situations, the user will not notice this difference at all. Performance could be a problem if your application requires large local data storage. For example, a sales force automation solution may require hundreds or thousands of contacts locally and the ability to sort them based on company name, contact name, or sales revenue generated. In those situations, you may notice performance issues during the re-sorting of data or other computationally intensive operations.

Memory requirements. RAD tools require runtime modules. Runtime modules are a layer between the application and the operating system of the device (see Figure 8.6). Three examples of runtimes are shown. This runtime layer performs many functions, one of which is a translation of the written application into something the operating system can handle. This layer can eat up anywhere from 50K to 1MB of space on the device. Memory on handheld devices is still constrained, and the installation time of the application onto the device can take much longer with a runtime.

Expandability. Many handheld applications require access to specific hardware peripherals or user interface elements not native to the RAD tool to complete their tasks. These include wireless modems for the transfer of data, GPS systems for location coordinates, barcode scanners for barcode reading, magnetic stripe readers for credit card

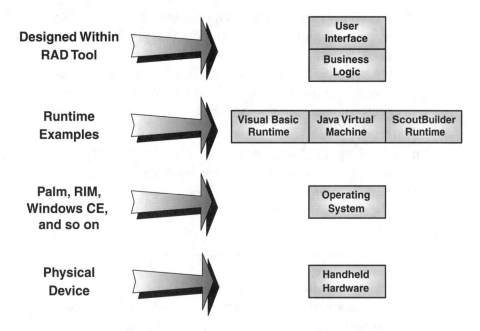

Figure 8.6 The layers of a handheld application written with a RAD tool.

transactions, and various complex interface controls. Some RAD tools enable for the creation of extensions to control these types of devices that you can't natively use with the RAD tool. Developers can write C or C++ extensions to handles things that the RAD tool cannot. This effort takes time, and some tools have better support for extensions than others. Don't get caught creating a world-class inventory system and find out you can't interface to the barcode scanner you need.

Proprietary nature. Many programming languages that RAD tools rely on are easier to use than C, C++, or Java. They may use a BASIC-like syntax. Because the implementation of this language will vary between packages and does not rely on some particular standard, this can introduce quirks and idiosyncrasies that you will not find in traditional languages. Ramp-up time to learn the language will take longer than standard languages.

A second area of potential issue is that RAD tools sometimes save their data on the handheld device in a proprietary format. If your enterprise solution relies on wireless middleware (and it likely will), ensure that this format is readable by the wireless middleware.

Our company has had great success with RAD tools with our clients. They are excellent for delivering demos, prototypes, and full-fledged handheld applications. They perform well in most situations and offer enough flexibility to solve a number of business issues. When you keep in mind some of the previous potential issues, you may still find certain applications that are better left to traditional development tools (C/C++).

Business impact. Many enterprise applications will succeed with RAD tools. Their ability to quickly develop solutions can cut months off of development time.

Java

Java, which also creates native applications, represents another opportunity for enterprises to leverage technology across different platforms and hardware. Its promise has always been that you would be able to write your application once and then run it on any type of computer. That promise is being extended to handheld and wireless devices in a variety of ways.

Sun's Java platform spans desktop computers, laptops, servers, handhelds, cellular phones, and other embedded devices. Anywhere Sun can find traction with its technology, they're going after it. The Java set of technologies has extensive breadth and depth. Hundreds of books, Web sites, magazine articles, and white papers are on the subject. However, it is worth discussing what Java for handheld devices looks like and whether or not your company can and should leverage it.

Configurations

Sun has created a version of Java in addition to their standard and enterprise editions that address the special needs of handheld, wireless, and embedded computing. Embedded computing involves the integration of Java technology onto computer chips that are installed in different devices. Examples include a car radio or a consumer refrigerator to control its operations. The version for these types of devices is called Java 2 Platform, Micro Edition (J2ME).

Instead of attempting to create Java implementations for every type of device out there, Sun created two generic configurations to address small form factor devices and embedded systems. The reason is that they believe most cell phones, for example, have similar processing power, screen capabilities, and input methods, independent of the actual implementation. However, enough differences exist that Sun addressed them through configuration and profiles. The two configurations are

Connected Limited Device Configuration (CLDC). Devices that fall in this category have very limited resources and computing power. Palm and RIM fall in this category currently. A CLDC device is *usually* a 16-bit device and has 512KB or memory or less. Most Palms, if not all, have more than 512KB of memory. However, if you compared the performance of a Palm today with Compaq iPaq, a CDC device, you would see an order of magnitude performance difference.

Connected Device Configuration (CDC). The CDC covers devices that have more power 32-bit processors and a minimum of 2MB of memory. In fact, the differences between the J2ME and Java 2 Standard Edition (J2SE, runs on desktop computers) are very little. Many applications with the requirements for the J2SE should also run on the CDC configuration of the J2Me. The CDC has full Java language support.

Profiles

Even with these standardized configurations, many manufacturers of devices integrating Java will likely find that these configurations are too generic to serve the specific needs of their devices. Sun then enables for the addition of *profiles* to a configuration. Profiles run on top of the configuration and can add functionality that is needed for a specific industry or type of device.

The first profile Sun released was the Mobile Information Device Profile (MIDP) for pagers and cell phones. The combination of the functionality of this profile along with the CLDC configuration defines how Java will operate on these types of devices. It dictates, among other things, the type of user interfaces you can build. If you consider that the MIDP was made generically for cell phones and pagers with integrated Java virtual machines, it's intuitive that the interface elements for building applications will be generic and limiting as well. This is not a criticism; it is simply the price to be paid for cross-device and cross-platform compatibility of applications. In the end, the enterprise should win with cost savings. For more information, visit http://java.sun.com/j2me.

Limitations

At various Java conferences, Sun demonstrated the ability to beam a Java application from a cell phone to a Palm-based device and have the same application run on both without modification. Impressive. However, a number of imposed limitations exist at this time with the Java Micro Edition. Many pieces have yet to be finalized or have just been finalized for the CDC and CLDC configuration.

At the time of the writing of this book, the number of classes or thoroughness of J2ME API is limited. For developing Palm applications, the number and types of controls available is extremely limited. Also, because the CLDC configuration (the one Palm uses) does not support the Java Native Interface (JNI), you cannot call or make use of the operating system directly. From an application standpoint, this cripples the range of applications you can create. If your application requires a wireless modem or GPS system, someone would need to make a Java class to control that device. Little support exists for supporting peripherals from Java on the Palm. Outside of basic data collection applications, the J2ME for Palm has little value to enterprises at this time.

Look to Sun to continue the evolution of its implementation of J2ME to support external devices, more user interface controls, and performance. J2ME is stable enough to being an evaluation, but likely not robust enough to build many of your company's applications.

Third Party Virtual Machines (VMs)

Virtual machines (VMs), or specifically, the Java Virtual Machine, are pieces of code that run on any computer capable of executing Java programs. In the case of the J2ME CLDC configuration, the VM is referred to as the KVM or kilobyte Virtual Machine because it implements its capabilities in the limited resources of small devices. See Table 8.2 for a list of Java VM manufacturers. The VM is responsible for translating the Java application code into code that a specific type of computer runs. There are Java VMs for the PC, Mac, UNIX, Linux, as well as a variety of handheld devices and smart phones.

Why this explanation? Because other companies can and do make VMs that run Java programs. In essence, they are compatible or should be compatible with Sun's Java VMs. One such company attempting to corner the market on Java VMs for handheld and wireless devices is Kada Systems. They've created VMs for Palm, RIM, and PocketPC that go beyond the Sun specification. The Palm's, for example, implementation from Kada Systems is a full Java Standard Edition implementation—more in line with the VMs that run on desktop computers in terms of capability. These VMs support a wider range of general classes, user interface classes, and the ability to utilize peripheral and software features not directly part of the Java class set. If you are set on finding a cross-platform development tool and want to maximize code reuse and minimize time, a third-party Java implementation may work for you. However, take the time to do a thorough evaluation of features.

IBM also makes a Java VM for small form factor devices. Their VM is called J9 and runs on a variety of processor and device types. Both Kada and IBM claim that their VM implementation is faster at running Java code than Sun's own KVM. IBM published benchmarks on their Web site comparing themselves to Sun when running Java applications on Palm handhelds. IBM showed that they were anywhere from 30 percent to three times faster depending on the device and the application being tested.

Table 8.2 Companies that Target Java for Handheld Devices and Smart Phones

PRODUCT	COMPANY	URL
J2ME	Sun	www.java.sun.com/j2me
KadaVM	Kada Systems	www.kadasystems.com
Visual Age MicroEdition	IBM	www.embedded.oti.com/
CrEme	NSIcom	www.nsicom.com

Java offers huge *potential* to enterprises. This is obvious. In the not-too-distant future, enterprises will be capable of rolling out cross-platform applications without having to rewrite huge amounts of code. If creating a Windows CE version of an application requires effort of level X, it may be possible to create Palm, RIM, and smart phone versions for a combined effort of 2X. Our recommendation is to keep your finger on the pulse of Java for viability in building applications for handheld and wireless devices. It will catch on as a mainstream opportunity in the near future, if it hasn't already. Some of the previous limitations with Sun's version hamper it from general purpose use in the enterprise in a large number of situations.

In Depth: Virtual Machines

Think of a VM as an imaginary computer—hence the term *virtual*. This computer has instructions that can control its operation just as a real computer does. There exist commands to write text on the screen, draw graphs, handle menus, access databases, and perform business calculations among others. The Java development environment creates programs that run on a virtual machine called the Java VM.

Sun has developed Java VMs that run on different types of operating systems: Windows, Mac, Linux, UNIX, Palm, and so on. They translate the instructions in the Java program to instructions for a specific operating system. For example, in Figure 8.7, the Java program at the top contains two commands in this simple program. The Java VMs for Windows and Macintosh translate those commands into ones that the specific operating system can understand. Notice the commands aren't necessarily the same, nor do they map one-for-one to the Java commands. However, the end result onscreen should be the same. This cross-platform capability is one of the largest advantages of Java.

This has two large implications. First, you should be able to create one Java application and run it on multiple types of computers. Second, the added layer of translating from Java's program code to something a specific platform can understand slows overall performance.

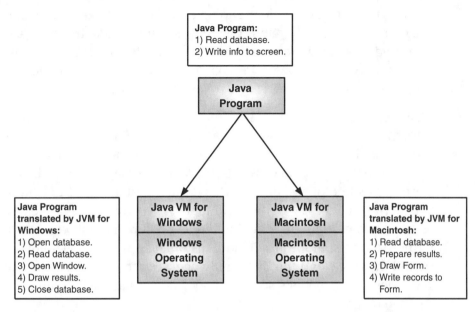

Figure 8.7 How Java VMs helps Java run across many different platforms.

Let's Talk about WAP

The Wireless Application Protocol (WAP). It's loved. It's hated. It's talked about to no end. It's also a Web-application-based technology. By sheer momentum, it has become a subject that the enterprise has hooked onto. The WAP can be thought of as HTML's little, but refined brother. WAP has been designed to facilitate Web-like content on small for factor devices. If you consider the limitations of handheld devices and smart phones today,

- Small screens (usually black and white)
- Limited memory
- Low performance processors
- Limited battery life
- Minimal input capabilities (compared to a keyboard and mouse)

you come to a quick understanding that HTML-based Web technology just won't work. Although HTML pages can be made that only use text, certain fonts, and black and white graphics, the WAP Forum saw the opportunity to tailor a markup language and related technologies to address the specific needs of wireless computing.

A consortium of companies—Motorola, Ericsson, Unwired Planet (now Openwave), and Nokia—got together to create a set of technologies for delivering Internet and Web-based content to wireless devices with limited resources. Instead of waiting for one company to introduce a proprietary technology to do this, this group created the WAP Forum, made the technology an open standard (if you have the money to pay the dues to participate), and invited any company that wanted to contribute to the standard to do so. This open approach has helped WAP avoid the tragic end that inevitably occurred to other less open technologies. Table 8.3 lists some of major players in WAP creating tools for its deployment.

WAP, now in its second full revision, is being met with limited, but growing success. The original 1.X versions, popularized in the media, had extremely limited functionality. This functionality was suited for simple tasks like showing stock quotes, but too weak to create enterprise strength applications. The founders of WAP did consider the use of Web technologies like HTML, Java, and JavaScript, but determined that they required too many resources to successfully implement on devices like smart phones and Palm handhelds. Outside of the basic presentation of data, WAP took into account some of the dangers of wireless computing like lost data due to interference and eavesdrop-

Table 8.3 WAP Technology Manufacturers

COMPANY	TECHNOLOGY	URL
Nokia	Servers, Gateways, Development Kits	www.nokia.com/wap/products.html
Ericsson	Servers, Gateways, Development Kits	www.ericsson.com/wap
Openwave	Servers, Gateways, Development Kits	www.openwave.com
Motorola	Servers, Gateways, Development Kits	http://developers.motorola.com/developers/wireless/products/

ping, and the relatively high cost of airtime when transferring data wirelessly.

Some of the pieces and features of WAP are

Markup language. Like HTML, WAP has its own text-based markup language for creating Web-like pages. WAP supports a much smaller subset of markup features than HTML due to the constraints on devices like cellular phones.

Scripting language. The closest equivalent to this in traditional Web world is JavaScript. WAP calls its language WML Script. It's responsible for enabling business and programming logic to be included with WAP pages on a device.

WAP gateway. Because WAP users will be operating a wireless device, some piece of software and hardware needs to sit between the WAP server and the WAP client translating the requests and responses between wireless networks and the Internet. This is one of the jobs of a WAP gateway.

WAP server. Listens for requests from WAP clients and responds to them with WAP pages.

WAP microbrowser. A Web-like browser that interprets WAP pages and shows them on a WAP device's screen.

Security. WAP includes a security layer to guard against data theft, eavesdropping, and modification.

Efficient data transfer. Due to the nature of the way WAP was designed, the amount of wireless traffic generated from a page request is much less than by the same application on an HTML browser over a wireless connection. The interaction between the browser and the back end is more efficient.

A WAP Transaction Step-by-Step

Let's look at an example WAP request and response to complete our understanding of the WAP architecture. As you look at Figure 8.8, you'll notice how similar the request-response paradigm is to a standard HTML request. WAP's architecture isn't much different except for the WAP gateway sitting between the smart phone (or handheld device) and the WAP server.

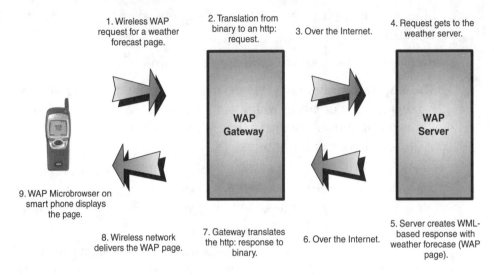

Figure 8.8   An example WAP transaction.

1. A user inputs a zip code within a WAP page on his or her cell phone requesting the weather forecast in his or her area. The smart phone encrypts and compacts this request before it is sent optimized for transfer over a wireless network like Sprint, Verizon, or Cingular.

2. The WAP gateway, residing at his or her service provider's network center, intercepts the request, decrypts, decodes it, and translates it to a standard http: request to be sent over the Internet.

3. The http: travels over the Internet to the WAP server that holds weather information.

4. The WAP server receives the request.

5. The WAP server creates a WAP page (using WML, described previously) containing the weather forecast information.

6. The WAP page is sent back to the WAP gateway over the Internet.

7. The WAP gateway encodes and encrypts the WAP response page for transfer over a wireless network.

8. The wireless network then delivers the page to the WAP-enabled smart phone.

9. The microbrowser displays the WAP page showing the weather forecast to the requestor. Forecast: Sunny and 80.

WAP: What Is It Good For?

An inordinate number of technology companies have WAP-based solutions they are selling into enterprise, and there is no shortage of magazine articles and books written on the topic. The reality is that WAP 1.X solutions are ill-equipped to handle the delivery of enterprise solutions to mobile workers. Typically, WAP is associated with smart phones, though WAP browsers exist for Palm, Windows CE, and other handheld devices. To date, smart phones have small black and white screens with few exceptions. Input is also limited to a small numeric keypad and a few extra buttons. WAP is useful in displaying simple information like contact phone numbers and meeting notifications and not much else.

It would be difficult (I'm being kind here) to deliver sales force automation, ordering, and healthcare applications with such a technology. Imagine trying to read and reply to an e-mail on a device that can show three lines of text with 12 characters on each line. If you had access to a larger screen and a more powerful device, it may be possible to throw a true Web browser (HTML) on the device. As higher speed wireless networks begin to roll out across Europe, Asia, Africa, North America, and South America, a true Web solution will be possible.

WAP 2.0, however, at the time of this writing is being finalized. Its specification was made in conjunction with the World Wide Web Consortium's (W3C) replacement for HTML, XHTML. Specifically, WAP is to become XHTML Basic. Without going into any detail, WAP 2.0 looks to be much more robust in terms of features than the WAP 1.x standard. It remains to be seen whether enterprise solutions can successfully be delivered with this technology. If you want to explore WAP or Web technology for enterprise deployment, look at WAP 2.0 (XHTML Basic).

As you begin thinking about wireless solutions for your enterprise, keep in mind device size and appropriateness. Even if WAP or XHTML had the most perfect design and implementation of any technology in the history of the world, a smart phone with a small screen will never be the way to collect field data. Something is inherently good and usable about a device with a larger screen and an operating system. Make sure you find devices appropriate to business processes. Never decide haphazardly to simply throw out laptops in favor of data-enabled cell phones. Each of these devices will serve a particular purpose. However, I believe you will find occasions where certain types of workers can work optimally with a handheld device in place of a laptop.

Security

Security on handheld devices is one of my largest pet peeves. Maybe I should say *lack* of security on handheld devices is my pet peeve. It more or less does not exist. Most devices, whether cellular phones or handheld devices, have some capability to lock the device from unauthorized use. To gain access, a user must enter a password before the device can be used or synchronized. If that person forgets the password, his only recourse is a hard reset of the device, which destroys all the data that was stored on it.

This type of protection also poses some business process issues. Having to enter a password each time the device is turned on is just flat annoying. How can you expect any type of worker who needs regular access to data to deal with this? Because most devices shut themselves off after a few minutes to save battery power, it would be unbearable to ask a user to enter a username and password each time he accessed the device. In the *Wireless Technology Trends* chapter, biometrics is covered. A biometric device would be an ideal way to authenticate a user of the device to the device quickly and easily. For example, each time a doctor needed access to patient information, she would simply put her thumb on a biometric reader and it would let her in. Today, a system like that requires a fair amount of processing power, which most devices don't possess, and is likely to be unrealistic in terms of cost. I expect that over the next 12 months that biometric system will be available for handheld devices in vertical markets that require such security.

Data Protection

Security experts will tell you (and I also believe) that a password or a biometric protecting the device is still not enough. It would be possible, in some cases, to remove the memory chip or card from such a device and use it with another. All of the major platforms currently store their data in an unencrypted format. When you think about doctors carrying devices around to facilitate working with patients, it seems ludicrous that all patient information would be stored this way.

Encryption is the process of using mathematical transformations on data to render it unreadable or understandable. Only someone with the key to reverse the process would be able to read the results. Depending on the strength of encryption, the ability to discover the key could take all of the world's computing power the length of the universe (in time) to decode. Realistically, encryption on handheld devices won't be this strong. However, it would deter most thieves from even trying.

Adding encryption so as to protect your corporate data can be done in three different ways:

- Purchase devices where encryption of stored data is handled through the operating system. To my knowledge, no major manufacturer does this yet. However, I believe it will be standard issue in the coming months and years.

- Purchase licenses to incorporate encryption libraries into your software development. Companies like Certicom and RSA sell them. Once you do, you can add this functionality to your applications. Before each data read and write, you would need to use the libraries to decrypt and encrypt the data, respectively.

- Purchase an add-on application that runs as a background task encrypting and decrypting data for your applications on-the-fly. Other variations enable the user to select individual databases for encryption and decryption. Although not an ideal solution, it does provide the best one for enterprises that require *immediate* and *low-cost* encryption. Companies in this space include JAWZ Inc, Asynchrony, Trust Digital, and Soft Winter.

General App Design

Whatever technology or technologies an enterprise chooses for handheld application development, design guidelines for those applications must be taken into account. Creating applications for handheld devices is not at all like developing applications for desktop or laptop computers. The limitations in handheld devices and current wireless networks prevent this. Some guidelines to keep in mind

Amount of data transferred. Wireless networks are notoriously slow. Unless you are in Japan or select parts of Europe, networks speeds hover at one-seventh to one-third of a 56K modem attached to a computer. Bypass graphics and sound and focus on getting the corporate data you need back and forth. These slower wireless networks do a fairly good job at information like stock quotes, sales histories, contact information, and so on.

Where possible, utilize the memory and storage of the device. If your device relies on graphics or fairly static contact information, preload it on the device and have the application take advantage of that local data store. Depending on the handheld device you're using, you may have a minute amount of memory or several megabytes.

Input methods. The user input methods for handheld and wireless devices fall into one of a few categories. Depending on what you are trying to accomplish, you will want to choose a particular device for rollout. For example, RIM has an extremely popular and powerful email platform. Its lack of a touch-sensitive screen may limit the types of general purpose enterprise applications that can be developed for it. Some of the most popular input methods for a user include:

- Touch screens
- Handwriting recognition
- Hardware buttons
- Hardware keypad (smart phone)
- Jog or thumb dials
- Expansion slot (serial port, SD, Springboard, and so on)
- External keyboards

Keep it simple. Handheld devices are here to make the average worker's job easier and more productive. A worker who has done his business process on paper for years should not be given a device that feels anymore complicated than the paper he gave up. To accomplish this, design applications that aren't as full-featured as their desktop siblings. Figure out what data and functionality is needed on the handheld and give just that.

This type of thinking is new when you compare it to desktop business applications. It is not uncommon to see software suites on computers to start at one size, and two versions later, it's five times larger and has 10 times the number of features. Don't burden workers with frivolous features, additional and unneeded options, and complicated interfaces. Rather, keep interfaces clean and simple, give them cues as to what to do next in the application, and keep the feature list to the exact set of options that are needed to accomplish the task.

Handhelds are not full-featured computers. This is one of my largest complaints with some peoples' views of handheld devices. I once heard it said that a Palm handheld will be a worthwhile device once it gets a full color screen, a miniature hard drive, 64MB of memory, and a keyboard. We already have those; they are called laptops. If you have the need for laptops in your organization, purchase them.

Handheld devices are incredibly portable and can make crucial time and location-sensitive information available. Likely the data you need on the device is also accessible for a corporate desktop computer. Do not plan on recreating that entire application on a handheld. That is not the intent of such devices. Select the necessary, highest value information and extend that to the device. It takes time to free your mind from traditional desktop design and development. It might take a couple of iterations to get to the feature set that enthralls the user and makes their work process much easier.

Free form text. This is a fairly specific design guideline, albeit a crucial one. Some business processes require the input of large quantities of free form text. *Free form* implies that what is collected can't be distilled down to a series of checkboxes or list items. The subjective nature of the data being collected requires large quantities of text entry. Entering data of this nature on a device, even with handwriting recognition can be a daunting or frustrating task. It may be necessary to choose a device with an integrated keyboard to make typing easier, add an external keyboard, or consider voice recording and transcription at a later time (see Figure 8.9). The final alternative and one we've had to recommend to clients at certain times is to not transform that particular work process into a handheld based one.

Write once for wireless networks. Each wireless network out there has a different set of functions that enable you to write applications that run over them. If you write the application (native) for Cingular's network, it is likely you'll have to modify the application to run over Sprint's network, for example. Many companies have software solutions that enable you to write your wireless application once and run it over a multitude of networks. Look at companies like Broadbeam and Aether Systems. However, browser-based applications will likely

Figure 8.9 Use a handheld with a keyboard for free form text input.

run over any network because most networks supports HTML and/or WAP browsers.

The Bridge from Here to There

This chapter covered quite a bit of ground. It discussed the different ways applications can be designed and developed. It compared Web-based application to what we termed native applications. We also covered the relevance of Java, RAD tools, and WAP to handheld devices. Finally, we looked at some general guidelines for the design of handheld applications.

As you can see, whipping up a simple application for a handheld, wireless device, or smart phone takes quite a bit of forethought. The interface, processing power, size, and development environments of these devices require making serious design choices. Many companies will have success or failure early on simply from the design of their application being well-received or not. Designing for small form-factor devices is 50-percent science and 50-percent art. Thrown together applications will not get used; don't fool yourself. Clever, creative, and intuitive applications will be popular with employees and customers. Prove it to yourself.

Think of a category of application like card games, personal information management, or project management. Go to Handango (www.handango.com) or ZD Net (www.zdnet.com) and download two or three competitive products. You'll quickly be able to separate winners from losers by examining the execution of application design in its interface, performance, and data display.

We started this book with the business of wireless and handheld computing and then took a several chapter dive into the underlying technology. Now that you may be all fired up to create your wireless strategy, design the architecture, and begin development, you're probably wondering what all this costs. Does a plant maintenance application cost $10,000 or $10,000,000? How much does true wireless connectivity impact cost? Those answers and more are in the next chapter.

CHAPTER 9

The Cost to Go Wireless

Information technology can sometimes be a funny area for enterprises. Measuring the total cost of ownership (TCO) for desktop or laptop computers can be a painful exercise in futility. Obviously, the capital and expense outlay for the hardware and software is easy enough to measure. It's probably not too much of a stretch to determine the cost of a technician to do an install of that hardware and software. However, how do you measure the cost of support phone calls, wasted productivity when applications crash, time spent walking to and from a jammed printer, and so on?

This chapter deals with the concrete aspects of wireless and handheld computing: the capital and expense portions of software and hardware, the cost of consulting services, and the amount of time it will take. TCO for handheld and wireless solutions has similar and not easily quantifiable costs, the same as desktop computing does. Every research group out there from Gartner to IDC has its take on TCO and how to measure it. This chapter does not attempt to tackle the esoteric sub-points of TCO because they are easily challenged and disagreed upon. Rest assured, though, they are there.

Cost Factors

Every client's number one question is how much does it cost? I can tell you this. The cost of the device does not reflect the size and cost of the project effort. Certainly handheld costs affect the overall price, but delivering wireless data to any mobile worker using some type of computer has a minimal cost infrastructure associated with it. We'll look at the major cost aspects of wireless and handheld computing in the following areas:

- Handheld devices
- Packaged software
- Peripherals
- Networking costs
- Wireless airtime
- Wireless middleware
- Middleware hardware
- Security
- Modifications to back-end systems
- Application development environments
- Project labor and expertise

Handheld Devices

The cost for a handheld device or smart phone has changed dramatically over the past couple of years. In 1996, Apple's Newton MessagePad sold for $799 and had few peripherals and little in the way of connectivity. The device, though elegant in design, was a bit ahead of its time. Today a wide range of devices exist at much more reasonable price points (see Table 9.1).

As you can see, the majority of handheld devices fall in the $200 to $500 range. Palm has more devices on the lower end of the range, while RIM and Windows CE fall on the higher end. Devices purchased through distributors and online stores typically offer some discount off retail. However, large quantity purchases don't produce as large of a discount

Table 9.1   The Cost of Handheld Devices

DEVICE	MANUFACTURER	RETAIL PRICE	NICHE
m105	Palm	$149	Palm's entry-level device
m505	Palm	$449	Palm's extra-thin, color, expandable handheld
Visor Deluxe	Handspring	$249	Handspring's mid-level Palm device
iPaq H3670	Compaq	$699	The multimedia Cadillac of Windows CE handhelds
SPT2746	Symbol	$1,995	Ruggedized, WLAN-enabled, barcode scanner-enabled Windows CE handheld
QCP 6035	Kyocera	$599	CMDA phone/Palm combination device
Blackberry 957	RIM	$499	Wireless handheld that excels at corporate email
Smartphones	Various	Free-$1,000	Many cell phones have WAP browsers built in for data capability.

as you might think. Margins for devices, especially for the lower-priced ones, are thin.

Many enterprises in vertical markets or with handheld needs on the shop floor, in warehouses, and on loading docks, opt for devices with better protection. Ruggedized handhelds have been around for a number of years from companies like Intermec, Symbol, and Telxon. The devices ranged in price, but many sold in the $3,000 range and ran old or proprietary operating systems. Many of Symbol's ruggedized handhelds are being offered with mainstream Windows CE and Palm operating systems (see Figure 9.1). The 1700 and 2700 series run the Palm and Windows CE platform, respectively. The devices are water-and dust-resistant, have integrated barcode scanners, and can survive repeated drops onto concrete. You can also purchase versions that have WLAN or WWAN technology integrated into them and magnetic stripe readers for processing credit card transactions and reading IDs.

As you can see from Table 9.1, the purchase price of such devices is much greater than standard devices. However, if you have any sort of

Figure 9.1 The Symbol SPT-x7xx handheld.

application where the device is used in a setting where it could be easily dropped, soaked, or crushed, the toughness of such a device will easily pay for itself. Finally, Symbol sells nonwireless handhelds with barcode scanners for under $1,000. Also, margins on these types of devices are higher than for traditional handhelds. Expect to be able to get better pricing.

Looking out over the next couple of years, the prices of handhelds will continue to drop. However, don't expect mainstream devices with numerous features to fall in the $100 range. Many features can still be integrated into handhelds as they mature (such as cellular phones, WLAN and WWAN technology, higher resolution screens, and color); consumers and enterprise customers should expect to continue paying in the same price range as today for these leading-edge devices.

Packaged Software

In emerging technology fields, the first set of wares any enterprise can purchase tends to be developer or toolkit-based. If you need a transportation and logistics solution, for example, you must build it yourself.

Unfortunately, this takes time, money, and in some cases, results in an unreliable solution because of the newness of the technology. As technologies mature, many companies will offer complete solutions with some pieces running on the handheld and others encompassing the wireless middleware. Aether Systems is one company taking this approach. They offer package solutions in a number of vertical markets. In the case of transportation and logistics, they've introduced one Mobile Max$_2$ for messaging and communication between trucks and dispatch centers. If solutions like these meet the enterprise needs, expect an up-front license cost in addition to integration costs. The overall solution is less expensive than one that is designed from the ground up.

Peripherals

One of the largest changes to handheld computing over the past 12 to 18 months has to be the addition of expansion slots to devices. Initially, expansion meant a simple modem that clipped onto the serial port of the device giving the ability for rudimentary connectivity. Others adapted wireless modems in the same way. Today handhelds support a wide range of standard connection ports including Compact Flash, Secure Digital, Springboard, and MultiMedia Card.

Whichever standard your handheld utilizes, you'll find any number of peripherals to expand the abilities of the device. Some expansion peripherals have a lot of business sense; others are purely for consumer's entertainment. Typically, a handheld device supports one peripheral connected to it. Table 9.2 compiles some of the common business peripheral categories with price ranges.

Table 9.2   Peripherals to Expand the Capability of Handheld Devices

PERIPHERAL CATEGORY	COST RANGE
Wireline modems	$100–$200
WWAN cards	$100–$400
WLAN cards	$150–$300
Global positioning systems	$150–$350
Digital cameras	$100–$200
Barcode scanners	$150–$200

Keep in mind that most devices can only accommodate a single periph-eral attached to it. If you know your solution will require wireless WAN connectivity and bar coding, choose devices that have one of them built in. This will keep any expansion slots available for secondary devices. Also, many of these peripherals can significantly increase the cost of a handheld. Although still much less expensive than a laptop or desktop to purchase and support, a wireless handheld device for your sales force can easily set the company back $800 to $1,000 plus a recurring monthly airtime charge. In the end, it's a small price to pay for the pro-ductivity and reduced costs through using enterprise applications that redefine corporate business processes.

Networking Costs

When you take your favorite Windows CE device out of its box, you get not only the device, but also the cradle. The cradle attaches to any PC (and sometimes a Macintosh) and lets you synchronize the data on your handheld with PC and corporate systems in some cases. As your enterprise moves to robust handheld and synchronization solutions, you will need to consider a plethora of networks beyond the standard cradle to get the data from the device to your company's back-end data systems.

Wireless WAN technology is one option, and the costs associated with it are different than those that run inside a company. Wireless WAN technology is discussed in the following section. It does not require any infrastructure outside the wireless modem. The network is provided by a variety of telecommunication companies more than willing to take your company's money in return for airtime. However, within the hal-lowed walls of any corporation, a number of methods are used to con-nect back to corporate data:

Network cradle. Network cradles look like a regular handheld cradle except they plug directly into a corporate network alleviating the need for an intermediate computer. This type of product is a low cost and effective way for a handheld device to access corporate systems. Price: $250 to $500 depending on options.

Network infrared ports. Network infrared ports are another popular alternative for those companies that find true WLAN technology too expensive at this time. This technology works in the same way as the

Ethernet cradle; it connects directly to a corporate network. The other end of the peripheral has an IR transceiver for sending and receiving data (see Figure 9.2). Most handheld devices with IR ports support data transfers and synchronization to network IR ports. So, they'll work with Palm, PocketPC, cell phones, laptops, and so on. Price: $200 to $250 per connection.

Wireless LAN technology. To incorporate wireless LAN technology (802.11b or Wi-Fi), you need two fundamental pieces: a wireless network card that connects to the device (if it's not already built in) and wireless access points that you hang from the ceiling. Price: $150 to $300 for the device that attaches to the handheld; $299 to $1,500 for an access point depending on speed, range, durability, and so on.

Bluetooth. Bluetooth may very well supplant Wi-Fi technology for handheld devices because it demands much less power from the device to operate. At the time of the writing of this book, Bluetooth devices are just now coming to market. Expect initial prices to be competitive with Wi-Fi technology. The hope is that Bluetooth chips will eventually drop under $5, enabling the technology to be embedded into everything from computers to cars and refrigerators.

Airtime

One cost that frequently is overlooked or just forgotten about in designing true wireless solutions is airtime. The pricing for time is in great

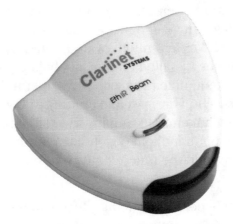

Figure 9.2 A Clarinet EthiIR Beam access port.

flux and will likely remain that way for some time due to competition between carriers and the constant upgrades to speed, bandwidth, and services.

Companies like Sprint, AT&T, Verizon, Cingular, and Arch Wireless have their own wireless cellular and paging networks. Many charge different rates either based on characters or kilobytes sent or minutes using the service. It is common to find flat rate pricing models as well, which allows you to send and receive as much data as you'd like. Excluding plans that limit the number of minutes or characters you can send per month, many of the plans fall into the $40 to $100/month range for service.

Intuitively, if you bring a large user base to a carrier, discounts on airtime will apply. Also, if your user base is scattered across multiple geographies, satisfying their need for wireless from one carrier will be difficult, if not impossible.

Some companies, like Aether Systems, have a Network Operation Center (NOC) that has data from multiple wireless networks flowing through it. They will host an enterprise application server, synchronization server, and give access to multiple wireless networks if you need those services. Additionally, they have the capability to negotiate pricing should your solution require different amounts of airtime from a multitude of carriers.

Wireless Middleware

Wireless middleware comes in so many flavors and fashions; it's hard to nail the price down. In general, companies use two pricing models: per seat and per CPU. Wireless middleware frequently requires that some piece of software be installed on the handheld device. That software is responsible for communication with the server independent of the transport method (IR, wireless, and cradle). Many companies charge a license fee for each handheld running the software. These prices range from $50 to $300 per handheld depending on the number of seats you buy, the company, and the functionality of the wireless middleware package.

The other model is per CPU pricing. Once you install the software on a server, you are free to connect as many clients to that server as you'd like. At some practical limit, performance will begin to degrade, and that will force your company to purchase either a faster server or a second server. Prices for these types of products hover in the tens of thousands of dollars range.

Both types of pricing models often require the purchase of a maintenance contract. Fees run anywhere from 10 percent to 35 percent of the original purchase price per year. This cost entitles you to technical support, upgrades to the latest version, and in many cases, training services.

Before you decide to go off and build your wireless middleware (many clients want to do that when they hear these prices), keep in mind the difficulty of creating your own. This software manages wired and wireless connections for multiple types of devices across many different connection types. Network programming is on the more difficult end of the easy/difficult spectrum of computer programming. Look at a practical example. Let's say the software costs $75,000 (1,000 licenses at $75/license). I challenge you to build a robust, multi-platform, scalable, multi-threaded synchronization or messaging engine for that same price. It just can't be done. Overall, this software is a small price to pay for the functionality and peace of mind you receive.

Middleware Hardware

Do not forget the related server hardware that is necessary for running this wireless middleware. Depending on the size of your organization, the number of users connecting to these middleware services, and your growth plans, you will need one or more servers of varying size. Also, many wireless middleware solutions are scalable with load-balancing hardware or software.

Judging the size of the server needed to support wireless and handheld computing is something of an art and a science. If your organization runs beefy servers (two or four processors, plenty of memory, fast hard disk systems, and so forth), you can expect to support 50 to 200 concurrent users. This number is dependent mainly on the amount of data that needs to be transferred or synchronized on an average basis. Some transactions may submit an order consisting of just one or two kilobytes of data; others may require that 1,000 item catalogs be installed on the handheld. As you prototype the application and develop the synchronization model, it will become evident how many users you can support on a server. Plan on installing separate hardware for wireless middleware and not using a server with shared services.

Security

Securing devices takes many different forms as we discussed in the chapter on security. If securing the data on the device through encryption is crucial, plan on spending tens of thousands to license encryption libraries from a security company. Other technologies companies have already licensed these libraries for their products, which do the encryption for you, and will license copies of these products for much less: typically $10 to $100 per handheld. These solutions encrypt and decrypt data transparently in the background as you access the data.

The different wireless carriers encrypt most airlink connections. That is part of what you pay for in your monthly airtime fees. Building in other types of security, like authentication, on the device adds a minimal cost to the development of solutions. Finally, although not prevalent today, a host of solutions for biometric security (thumbprint, retina scan, and so on) options are just now coming to market for handheld devices and will provide a reliable security platform for access to data on handhelds.

Modifications to Back-End Systems

Most companies already have developed the corporate systems they need. They want to use wireless technology to extend that data and those transaction engines out to mobile workers. However, these systems must be modified in order to support wireless. If you are simply pushing data out to a device, there is likely no additional cost. More times than not, corporations require the synchronization of data between a back-end system and one or more handhelds.

To support this model, the back-end systems must be modified to handle the detection of changes in data. See the chapter on wireless middleware for more details. In summary, if a corporate database cannot return record sets of new, deleted, and modified data to the wireless middleware, many of the advantages of synchronization will be lost. Plan on development time to modify these systems to handle the detection of changed data. It's hard to estimate that as a concrete cost because it depends on the database size, the extent of support for synchronization, and a number of other factors.

Application Development Environments

Your developers or consultants will need to build applications using one of many application environments. Some environments create native applications and others are RAD tools. Table 9.3 highlights some of the more popular packages available for each platform.

Countless tools from a number of manufacturers support the development of Web-based applications. In fact, there are simply too many to mention. I'll wager that a number of tools you have in-house will likely suffice for development of these applications.

Project Labor and Expertise

Before starting any enterprise project, it must be known where the labor and expertise for that project will come from. In addition, the project expertise must come in many different areas: strategy, process redesign, technology design, development, integration, implementation, and training. Many times companies decide to outsource some or all of the project areas listed previously. Why? Many reasons exist:

- They do not want to hire full time resources for a project lasting a short period of time.
- The technology is in a field where the company does not have expertise.
- The project is too critical to figure out for themselves.
- Resources to put against the project are lacking.
- The timeline is short.

Whatever your company's reasons are for looking outside, keep some things in mind. Some of these factors are related specifically to the technology, others to its maturity, and finally the third category is simply general guidelines for things you should expect from any first-class consulting firm:

Table 9.3 Development Environment Options and Costs

PACKAGE	MANUFACTURER	PLATFORM	LANGUAGE	PRICE	URL
Satellite Forms	Pumatech	Palm	Proprietary (BASIC-like)	$995	www.pumatech.com
ScoutBuilder	Aether Systems	Palm	Proprietary (BASIC-like)	$500	www.aethersystems.com
Codewarrior	Metrowerks	Palm	C/C++	$369	www.metrowerks.com
AppForge	AppForge	Palm	Visual Basic	$695 (Professional)	www.appforge.com
JavaVM	Kada Systems	Palm/Windows CE	Java	$895	www.kadasystems.com
Visual Studio for Embedded Devices	Microsoft	Windows CE	C, C++, VB	Free	www.microsoft.com
Visual Studio	Microsoft	RIM	C++	$979	www.microsoft.com
MobileBuilder	PenRight!	Palm/PocketPC	C	$1,595	www.penright.com

Price. Domestic development resources range in price from $50 to $200 per hour depending on the level of skill and technology involved. A number of small shops out there bid inexpensively on projects to get them. However, be leery of those that have not done actual enterprise engagements. Simply throwing a Satellite Forms or Visual Basic program on a device is a far cry from an end-to-end, secure, scalable, and wireless enterprise solution.

Development experience. Experience is key in an emerging technology field. If you plan to pay consulting fees, ask whom they have served. Ask what the projects were. Did the project impact one person, a department, a division, or the entire company? Does the firm have experience and understanding of the whole wireless architecture and all of the design choices, or do they simply push a particular technology from a particular vendor? Does their experience allow them to speak easily on the different platforms, connectivity options, wireless middleware choices, and the differences between them? Simply having wireless on a service line list does not qualify someone to do the work, especially in an emerging field.

Strategy experience. With a technology like wireless, companies have the opportunity to redefine how they do business. A consulting firm should have the capability to guide you through your wireless strategy and how it will impact your business. Does the firm have the strategy and business expertise to guide you through determining *what* to enable and *why* you should enable it in addition to *how* you should enable it? Simply said: The strategy and process redesign of how wireless and handheld technology will impact your employees, products, and services is the single largest factor in determining its success. Unfortunately, it's not a place many companies spend much time.

If a company has made the mental leap to a prototype or created a wireless solution, they've usually set aside funds and resources to tackle the technical aspects. Rarely do they have to spend time with such activities as use cases, focus groups, process redesign decisions, package evaluation and selection, market analysis, and how wireless can impact the company. That said, you also do not need to spend millions of dollars and affect every single worker. Spend time prototyping and beta testing. Use strategy sessions to make sure what you are doing keeps in mind what the company is trying to accomplish long-term.

Overseas development. I've heard both sides of the story here. One says that the development costs were incredibly inexpensive, and they saved a huge amounts of money. I've also heard that lack of communication, development teams being separated from each other, and language barriers have cost so many extra hours that it would have been the same price or less to have consultants work on-site. This is one area to possibly consider. Depending on the sensitivity of your data, it may not be feasible to utilize these resources.

Cost assignment. Some wireless projects involve strategy and process design only. Others are package-selection focused. Some use Web browsers and others use synchronization servers. For those projects that utilize wireless middleware in addition to client application design, strategy, security, and other pieces, one of the largest costs is in building the connectivity from the handheld to the back-end system using the wireless middleware. In fact, it is not uncommon to find that one-third to half of the development cost (consulting fees, if you outsource) is in the business logic of the middleware and the modification of back-end systems. This is typically the crux of the wireless solution because it implements business logic, integrates multiple systems, and processes transactions.

ASPs or In-House

Companies can take different approaches for supporting wireless middleware and application servers. Deciding to host everything at your company can require a significant time and cost investment. Setup and installation of servers, networking, fail-over hardware, and security services all add to the cost of wireless computing. In addition, your company must have people trained with the technology to support it. Because of this, a number of wireless software, hardware, and service companies offer an outsourced or Application Service Provider (ASP) approach. The exact definition of an ASP is nebulous and different depending on whom you ask. However, I am defining it as the outsourcing of specific infrastructure, server, and application pieces to a third party. These services are becoming popular with enterprises seeking to reduce overall cost by outsourcing this overhead to expert companies and instead paying monthly fees. Figure 9.3 shows a quick overall architecture of how this approach may work.

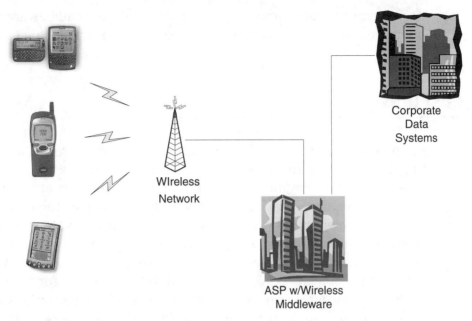

Corporate
Data
Systems

WIreless
Network

ASP w/Wireless
Middleware

Figure 9.3 Using an ASP for hosting wireless middleware.

Basically, when a user of a handheld device makes a request for information or synchronization, the wireless middleware is triggered. Instead of the middleware being located at the company's corporate headquarters, it resides at an offsite facility hosted by the wireless middleware provider. That ASP may host the corporate data onsite, or the ASP may make a request to your company's corporate data behind your firewall. Some obvious trade-offs exist with using an ASP instead of doing the work in-house:

- It's less expensive up front because your company will only need to pay a monthly fee (likely per user). It can be more expensive in the long-term because the fee is on-going.

- Less infrastructure and support personnel are needed at your company.

- Security could be compromised because the data goes through a third party's network.

- Monthly fees entitle you to the latest software.

As you explore your options, question your wireless middleware vendor about the ability to outsource and determine if it makes sense for your company.

Cost Recap

I don't know if the range of costs that can be incurred from a wireless product or service surprises you or if it's in line with what you expected. Either way, one the most frequent questions I get is how much does wireless cost? You are certainly educated enough now that you can honestly stand back and utter, "It depends," just like I would. However, you can now quantify the range of technology products and services that you need to consider. On one end of the spectrum, you can create real, wireless prototypes and demonstrations for thousands of dollars. You can create robust, albeit small, wireless solutions with proven ROI for tens of thousands of dollars. On the other end, large strategy and implementation projects involving thousands of workers, millions of consumers, and multiple geographies can push the budget into the hundreds of thousands or millions of dollars. In between is a heck of a lot of elbowroom. The important thing to take away is that *the cost of entry to wireless with a real ROI can be inexpensive.*

Certainly management buy-in for a wireless project is crucial for its success. One way to gain that buy-in is to start with a small, but beneficial project. Creating a proof-of-concept or prototype for use with focus groups or management teams that will ultimately fund projects is a proven winner for wireless. Many of our largest projects started with a proof-of-concept for an enterprise client. Showing what is possible and demonstrating the ROI first-hand opens many eyes to the wide range of possibilities. We demonstrated to one company how a wireless help desk system for tracking computer issues and assets greatly increased the productivity of their technicians. In another case, a demonstration of a handheld healthcare solution saved inordinate amounts of time in collecting patient information and reporting it to the government for insurance purposes. These types of applications took but a few weeks to design and develop.

One last thing: Let's also break down the costs we talked about in this chapter in terms of one-time and recurring costs. All costs are ultimately recurring, but I will differentiate based on expenditures you must shell out on some pre-defined time interval. Table 9.4 covers these costs. Please note that not all projects have all of these costs.

The Bridge from Here to There

All wireless projects have costs. Enterprise projects have a wider range to consider due to the enterprise-strength infrastructure that must frequently be considered. However, wireless projects, whether strategy or implementation, do not need to cost millions or even hundreds of thousands of dollars. You can start with a concept and build a prototype to prove your idea. If you're out of ideas, a number of talented firms exist with deep expertise in wireless on both the strategy and development side. Tap them for ideas and resources to help your firm identify the low-hanging fruit within your organization.

This chapter is one of the first that pulls away from the technology discussion and focuses on the business aspect of cost. Hopefully, this has given you insight into the different pieces of the wireless architecture

Table 9.4   Wireless Project Costs

ONE-TIME
Handheld devices
Packaged software
Peripherals
Networking costs
Wireless middleware
Middleware hardware
Security
Modifications to back-end systems
Application development environments
Project labor and expense
RECURRING
Airtime
ASP services
Handheld application and wireless middleware maintenance fees

and what to expect from a cost standpoint. We've now completed our tour of wireless strategy, architecture, and technology. The next two chapters look at real case studies of handheld/wireless enterprise implementations. The first is a strategy and development engagement dealing with ordering and inventory in the restaurant industry. Enjoy.

CHAPTER 10

Wireless Case Study: Ordering and Inventory

The first wireless case study is within the restaurant industry, specifically the independent restaurateur. This covers restaurants that are not chains, but may have a few locations at maximum. For those not familiar with this market, let's look at how much impact it has on our economy. Check out these stats:

- The overall impact to the U.S. economy in 2001 is expected to be $1 trillion. This includes everything from restaurant workers to food and equipment manufacturers and the transportation involved.
- Over 11.3 million people are employed in this industry in the United States.
- On a typical day, almost half of all adults (44 percent) eat in a restaurant.
- In 2001, purchases by restaurants for food and beverage products will top $142 billion alone.

(Source: National Restaurant Association)

With statistics like these, it is no wonder hundreds of technology companies offer products and services to the industry. This chapter covers a dot-com that entered the space to make restaurant operations more efficient for independent restaurant owners (not chains).

Industry Background

RestaurantB2B (not its real name) started life with millions of dollars in funding and set out to redefine how the $125 billion independent restaurateur industry went about its business. Eschewing the traditional PC-based application models, they set out to leverage web technology and a homegrown Business-To-Business (B2B) commerce engine to bring together restaurant owners and suppliers.

The restaurant industry is notorious for being low-tech, paper-based, highly competitive, and cumbersome. Many individuals who own and operate a restaurant independently do so not for the money but for social aspects of being able to serve returning customers. Unfortunately, it does not turn out that way for most owners. Due to the nature of the business, it is not uncommon for an owner to spend 60 to 80 hours a week running the restaurant, *every* week. This dot-com came into existence to help make the lives of independent restaurant owners easier.

Our client set about to reduce the time the owner spent running his or her business so he or she could spend more time with the customer or, more importantly, somewhere else doing something else. Web site access to RestaurantB2B.com is free to the restaurateur, with revenue being generated on each transaction for the goods and services being purchased through the site. This ultimately means the suppliers who signed on to distribute their goods pay for the service. In return, suppliers have access to new customers not in their geographic area.

Restaurant B2B sought to solve all types of issues with restaurants, not just serving as a transaction engine. Table 10.1 describes the range of services offered. Independent restaurateurs can't look to a corporate headquarters for policies and procedures as well as hiring staff. This dot-com offered these services to its clients instead.

Table 10.1 Services for the Independent Restaurateur

BUSINESS TOOLS
Creating income statements
Calculating ratios
Determining menu profitability
Handling restaurant open and close
MARKETPLACE
Finding and buying cookbooks
Finding and buying restaurant equipment
Asking restaurant experts and chefs advice
STAFFING
Recruiting employees including chefs
Hiring procedures
Discipline procedures
Keeping employees
ORDERING
Ordering food items

Ordering and Inventory Background

RestaurantB2B's largest area of assistance and expertise was on the ordering and inventory side of the house because numerous inefficiencies existed there. Restaurant owners must battle a number of challenges when creating orders for food items:

- Dissimilar order forms between vendors require learning the intricacies of each. Many times this forces the owner or an experienced manager to place the order using his or her time instead of a regular employee.

- Ordering requires many forms from multiple suppliers who ask for redundant data like name, address, and phone number. Filling out these forms takes inordinate amounts of time.

- Items that need to be ordered are scattered throughout the restaurant. The owner or manager must walk around with a clipboard in a somewhat random fashion collecting inventory and order information.

- The collected order information can easily be transcribed or interpreted incorrectly causing errors at different points in the process.

- Many times submitting orders requires phones calls to individual suppliers along with follow-up calls should the order be late or missing particular items.

To alleviate a number of these problems, RestaurantB2B created a unified ordering process. The owner of the restaurant logs into his or her account on the site and builds one or more catalogs for the items the restaurant needs. These items could be dry goods, meats, utensils, wall hangings, produce, and so on. The restaurateur also has access to substitute goods should a supplier be out of inventory or a competitive supplier have a substitute at a lower cost that meets the needs of the owner. Such a B2B ordering system could alleviate some of the previous issues in the following ways:

- Order forms (onscreen) are simplified. They have an identical look independent of the supplier delivering the products.

- Items onscreen can be ordered on a catalog page to match the layout of a restaurant. For example, an owner could group items together by location: pantry, freezer, behind the bar, and so on, and then by location on a shelf top-to-bottom and left-to-right. It is then easier to collect the information on paper by going to each location one time and not missing items.

- Orders are placed through the Web site. Owners are notified of delivery dates and delays through the Web site as well reducing the number of individual phone calls to suppliers.

However, not all of the issues are resolved through such a system.

If you think about it, the restaurateur is a mobile worker even within the confines of the restaurant. Ordering and inventory, the business process, is best performed on location. This gives this process location sensitivity. As we discussed early on in the book, countless business processes with location-sensitivity make great candidates for a handheld and/or wireless application. To finish reaping the benefits of an electronic procurement system, handheld technology needs to be added at the point the data is gathered. This will answer any remaining questions related to transcription errors and duplicate data entry.

Wireless Strategy

RestaurantB2B asked us to guide them through their wireless strategy and identify key areas that handheld and wireless technology could make restaurateurs' work easier. We found several candidates outside of ordering and inventory that made sense as well. We helped our client prioritize this list based on a number of factors. Ordering and inventory is obviously a hot area for this company that ended up having the great overall value to customers and is the focus of this chapter.

As we defined the strategy and technology architecture to support wireless computing, we needed to look at the average restaurant's infrastructure for handling a Web-based system with a handheld extension. In reality, few owners used a PC on their premises and if they did, there was only a small percentage chance that the PC had a connection to the Internet. The takeaway here, which is intuitive, is that restaurant owners by and large are not technologically sophisticated. There was never a pressing need to do so. Also, restaurants are scattered across the United States, which says something else about how connectivity for a handheld solution should work:

- Due to the unsophisticated computer systems restaurant owners operate, the likelihood a true wireless LAN at a restaurant is near zero at this time. This prevents one of the methods for having a true wireless solution: WLAN.

- Restaurants are indoor operations in the vast majority of cases. Due to this, you cannot be guaranteed that a cellular or two-way paging data network will have coverage within the restaurant. This prevents the other major method for creating a true wireless solution: WWAN.

- RestaurantB2B's system needs to operate efficiently over a 56K modem. If the restaurant owner somehow had a faster connection, it was a bonus. Also if the PC is the primary connection to the Internet, it is almost assured that it will also be the primary connection to the Internet for the handheld devices. Owners will synchronize and transfer data by dropping the handheld in the cradle and manually initiating the synch process.

- When data synchronization has to occur, there are two primary choices for where to install this code: the computer, or a handheld device that connects through or centrally located on a server. Due to the geography challenges of supporting restaurants all over the country and the need for a simple update process to synchronization code, an enterprise synchronization solution (a.k.a. wireless middleware) was needed and hosted at RestaurantB2B's location.

Platform

Platform decisions were made early on in the development process. In this case, RestaurantB2B chose a Palm-based solution. Two primary reasons drove the decision to adopt this platform:

Market share. During the summer of 2000, PocketPC (Windows CE Version 3.0) had recently been released and market share of these devices was a fraction of today's. Palm, on the other hand, held the lion's share. Devices could also be purchased from a variety of manufacturers: Palm, Handspring, IBM, Sony, or Symbol.

Cost. Windows CE-based devices are, on average, more expensive than Palm-based devices. Given the functionality of the ordering and inventory system (data entry and lookup), there were no compelling reasons to choose a more expensive device. Had our client needed some functionality exclusive to Windows CE like video, high quality audio, or serious processing power, we may have chosen differently.

Back-End System and Synchronization Requirements

The back end database of this Web site and related technology was also outside of the scope of our work. In other words, we were not allowed to make changes to it to support wireless devices. The back end system is based on Oracle with various Java layers and application servers. The front end to this system, the interface for the restaurant owners, is Web-based.

One of the main requirements for any handheld or wireless device connecting to this back end system is that the device accesses it through URLs (http: requests) as if it were a regular Web client. This guarantees a consistent interface for any type of client that connects to it. Figure 10.1 shows the various layers of business logic and code that transform the raw restaurant data in Oracle into something presentable and

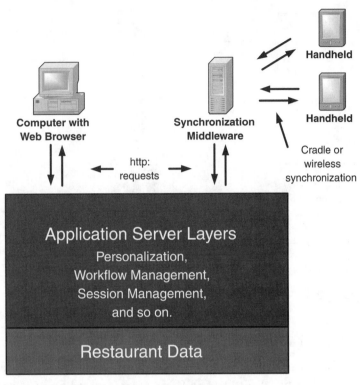

Figure 10.1 RestaurantB2B's back end architecture.

usable within a Web browser. Any device or other system that might access raw table information directly from Oracle would not gain the benefits of the business logic in the higher layers. More importantly, incorrectly updating data at the Oracle table level without using the previous layers would result in a corrupt database.

This fact has severe implications on selection of a wireless middleware package. Many of the server-based packages available today interact with databases directly on a table and field level. In fact, the paradigm is mapping a field on the handheld device with one inside of the database (see Figure 10.2). When a change occurs on either end, it is easy for the middleware to apply business logic rules to handle these conflicts. For example, let's say the restaurant owner uses the handheld to order five cases or pickles. The next day on the Web site, he orders seven cases of pickles (forgetting he recorded five cases on the

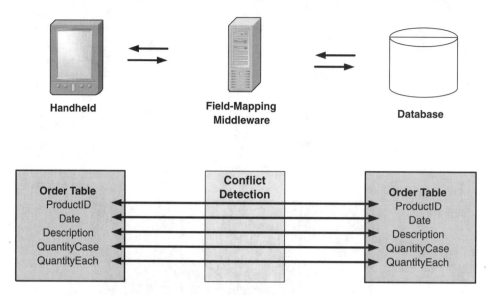

Figure 10.2 Field mapping wireless middleware.

handheld the day before). At some time later when he synchronizes the data on the handheld to the Web site, the field mapping middleware catches the changes on both sides and uses business logic to resolve the issue. If the rule stated that orders on the handheld override orders on the Web site, the Web site's order is updated to reflect five cases of pickles. A field-mapping methodology has the advantage that setting up the middleware to synchronize tends to be easy and straightforward. On the downside, it can in no way support synchronization the way we needed it to for the client because the requirement was to make http: requests to a Web server.

In the end (Design V2 in the following), we chose a wireless middleware package that supported a more generic synchronization model. It could connect to databases through a variety of methods; one of the ways it could was through http: requests instead of field mapping. Each http: request submitted collected data and each http: response included order guides, pricing information, and so on. The received data was parsed, placed into databases files a Palm handheld could interpret, and sent back to the device.

Design V1: Web-Based

The fun began on our very first meeting with the client. The day we set foot in B2Brestaurant's offices, we were asked to leverage a Web-based model for their ordering system. Their reasoning was that because they were Web-based and invested millions of dollars into Java technology, we should reuse that wherever possible. As we discussed in Chapter 8 on handheld application design, solutions with handheld applications that require heavy user input and data manipulation tend to fare poorly when created with Web technologies. We consulted with them (to no avail, initially) to use a native application that talked directly to their Web site for performance and interface reasons.

We proceeded with a Web-based solution and synchronization server. This synchronization server had the capability for synchronization groups of Web pages to handheld devices for use offline when not connected wirelessly. It also supported a JavaScript module for handling some of the business logic on the device. Although we found it easy to deliver Web content to the device, we simultaneously found it difficult to interact with this data for three main reasons:

Performance. Because a restaurant could potentially have hundreds of items in an order guide, we needed some way to logically show them on a handheld device other than a gigantic scrollable list. Items were split into categories and locations, which reasonably reduced the number of items to about 20 on a screen. Unfortunately, the Palm is ill-suited to handle this many items on a Web page and as a result, page navigation performance was poor.

Interface flexibility. As a result of a handheld small screen, clever and inventive ways to display data were needed. Our client was limited by the standard GUI controls of a Web form instead of what we ended up doing in V2 (in the following): creating a custom tree-control like Windows Explorer in Microsoft's operating systems to sort items hierarchically.

Saving state. Saving state refers to the process of storing data collected on one screen and making some or all of it available on the next screen. PC Web-based solutions also have difficulty with this because most rely on submitting the collected data to the Web server when switching pages. For example, if you enter a billing address on a Web page and want to make the information available on the next page as the default for the shipping address, the information must first go through the Web server. Few choices exist for storing the information locally while the Web page switches. Because this solution ran offline (no real time connection to the server), we had to add custom programming to be able to save data on the handheld as the Web pages switched.

Work on this prototype was rapid. We had two full-time resources from our company working on this solution along with one to two FTEs from our client supporting necessary changes to their back end to complete this work. Elapsed time was approximately six weeks of effort. Once a working prototype was complete, we along with RestaurantB2B's project manager, demonstrated the application for the CEO. Within just 10 minutes, he threw the whole thing out because performance on the handheld was poor and interface was inflexible. Even though six weeks was lost, it served as a great lesson on when to use a native application versus a Web-based one on handheld devices.

The Web-based solution for this client was a failure not because the technology is inherently bad, but simply what it was being asked to do. To be fair, the technology we used for Version 1 is a robust, scalable platform. It, unfortunately, was not suited to the rigors of the *type* of application we were implementing.

Design V2: Native Application

Immediately afterwards we began the design and construction of the same solution based on Metrowerks Codewarrior, which is Palm's C/C++ environment for creating native applications. Codewarrior gives the most performance of any development language for Palm currently. Due to the number of products, items, and ways in which the user wanted to sort them, the application demanded a high performance tool like Codewarrior. As anyone who has done C/C++ development knows, the trade-off for this performance is a longer, more detailed development cycle.

One of the largest design challenges for our client and us was the architecture of an application that supported the easy lookup of any one of hundreds of products. The Palm has available to it the ability to create lists and tables, but nothing optimized for categorizing data. For this client, we opted to create a custom control that sorted data with a tree structure in the same way Microsoft Windows Explorer does. Figure 10.3 shows a screenshot from Windows Explorer. On the left half of the image, notice how folders and subfolders are organized in a hierarchy. We adopted this methodology for our client. The tree control's design included the ability to sort the data in different ways. Whether the user decided to sort the data by location, supplier, or some other criteria, the hundreds of items were quickly condensed into manageable choices.

In the end, the basic functionality of the handheld application enables for restaurateurs to collect one or more orders on the device. For each order, they collect the quantity they have on-hand (for inventory purposes) and an order value. A prompt at the bottom of the screen reminds them of par value they would like to keep on hand. Par values, for those unfamiliar with the restaurant industry, are defined and set quantities of an item the restaurant would like on-hand at all times.

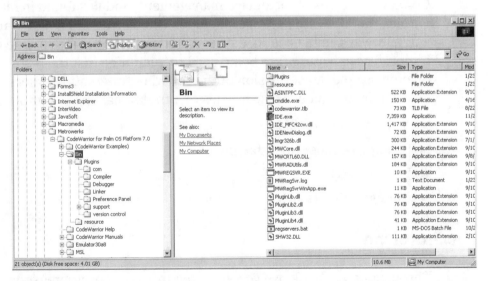

Figure 10.3 The second and final design uses a custom tree control similar to Windows Explorer.

Restaurants will set par value differently based on the week, the weekend, seasonally, or some other criteria. For example, if a restaurant has a seasonal special for a certain type of chicken dish, the par value for chicken breasts may be higher for the summer order guides. All orders can then be synchronized, but not placed, with the RestaurantB2B's Web site. The manager or owner needs to login to the Web site and verify the items, quantities, and pricing are correct before submitting. Orders could have been submitted from the handheld; our client chose not to allow this in the current version of the software due to the nature of the restaurant owners' business processes.

Synchronization occurs through wireless middleware that handles general-purpose business data instead of Web data. It also does not rely on a field-mapping paradigm. Again, the reason is that the middleware is not permitted to talk directly to the database tables. The middleware needs to make requests like a regular desktop Web client and have the capability to submit, accept, and parse Java objects.

Many of the wireless middleware packages we reviewed also had the ability to manage devices. Another requirement of our client is to be able to upgrade restaurant owners with the latest version of this application, other applications, and databases of information without shipping upgrade CDs to the owners each time there was a change. Common features of the device management modules found in many wireless middleware packages include device inventory of software and hardware, data backup, data restore, and electronic software distribution. These features are being used to guarantee that the restaurant owners use the latest revision of the software for creating orders. It also eliminates many of the hassles of a traditional upgrade process:

- Mailing out CD-ROMs
- Sending e-mails with attachments
- User errors
- Configuration issues

The Prototype

Throughout this book, we've advocated the use of prototypes to try out concepts and gain management buy-in. We also did this with RestaurantB2B. Between Design V1 and Design V2 existed V1.5. We created a mock-up of the order system with a RAD tool for three purposes. First, management needed to give their buy-in for the new system. Using a RAD tool allowed us to quickly get sign-off on designs. Second, we needed to rapidly start working with focus groups to get their input on features and screen layouts. Finally, we created a flashy prototype for use at trade shows and to show investors. It included basic desktop synchronization and other features like barcode scanning, receiving of goods, and alerts. This model of development enabled our client to filter and prioritize features for future versions of the real software.

Results and Benefits

The system and its successive iterations required about six months to get a solid first version in place. I would call this atypical for a design and implementation. Some of the tools available today, only 12 months later, are more robust and easier to use than the tool options available back then. Additionally, there were times where development of the mobile portion of the site was ahead of the development of the regular site simply because of the scope of work. The wireless portion required four to six FTEs, whereas 50 to 60 FTEs worked on the rest of the site. However, it is interesting to note that if we had to redesign the system today, we would likely make all the same decisions not only with the design, but the specific technology choices as well, which I believe is a testament to the relative stability of standards in the industry.

What We Got Right

Strategy. After listening to the management team tell us what they wanted, what their vision was for the wireless industry, we set off building it. We spent time gathering requirements from restaurant owners and chefs, the people who would ultimately use the system. We prioritized handheld and wireless ideas based on the ROI. We also wrote a lot of documentation.

Management buy-in. Management from the CEO on down believed that wireless and handheld technology would be crucial to their success. They had many (dozens) of ideas for wireless; some were more practical and beneficial than others were. However, buy-in from management allowed for capital and other resources to be allocated for wireless projects. Although the management team advocated the wireless aspects of the Web site, the team was not allowed to participate in design meetings, which was an oversight, not an intentional decision. See the following.

Handheld development. Using a native application on the handheld was the right choice then and it's the right choice for today. The performance of the application just can't be matched by a Web application at this time. One technology that may be feasible in the near-term that was not then is Java. Java gives the ability to support Palm and Windows CE with one application. Also, running Java in a Web browser over a high-speed wireless network may enable for Web applications that can compete with native applications in performance one day soon.

Wireless middleware. Using wireless middleware instead of a conduit on individual restaurants' workstations was a great choice by RestaurantB2B. Doing this permitted the ability to centralize synchronization and support, handle security issues, remotely update devices, and quickly make changes to conduits. Additionally, data transmissions are optimized. A 500-item product catalog can be synchronized to a handheld device in about one minute.

User requirements. Through use cases, focus groups, and talking with on-staff chefs and other personnel, the handheld application matched to a large extent the requirements of the users. As beta testing began with a small group, they requested few changes. Most found the application useable, intuitive, and more productive than the paper process.

What We Learned

Web versus native. Although we got it right in the end, we should have stood our ground. You always want to support your client and their wishes, but you must also act in their best interest even if your point of view stands in contrast in theirs. Based on that experience and others, we are now in a great position to give clear reasoning why one will perform better than the other. If you still can't convince them, a quick prototype demonstrating the limitations will usually remove all outstanding questions.

Wireless from the beginning. The development of our client's Web site was in full swing the day we arrived to help them. Therefore, we didn't have much say in the design. Ongoing, the wireless team (our staff and theirs) was kept in the loop, but we did not participate in many design meetings. At some level, they treated wireless as an extension of the site, not an integral part. For example, the site did not initially take into consideration the particular needs of handheld synchronization. To support synchronization, a back end system must be able to detect changes in data. In many databases when a record is deleted, it is gone from the system. If a record is deleted, but no history of that is available, it becomes impossible for the wireless middleware to recognize that change and synchronize it with the handheld.

We made a few changes along the way that affected tables, fields, queries, and objects within the database. Although it did not delay us significantly, it could have been avoided had wireless synchronization been considered in the original design. No blame is to be assigned. Wireless is a new technology and I believe many did not realize changes needed to occur to account for handheld technology.

The Bridge from Here to There

RestrauntB2B has experienced its share of issues along with many other dot-coms out there. However, their business model is sound and truth be told, the restaurant industry is decidedly antiquated. In numerous places, technology can improve and streamline operations. Unfortunately, these efficiencies will come at a more paced adoption rate and cannot be forced. Wireless/handheld ordering and inventory systems are one step in the right direction. As RestaurantB2B continues on, no doubt they'll find other places for this new technology.

This chapter's case study looked at where the rubber meets the road in wireless and handheld computing. We'll continue this in the next chapter. We'll shift gears a bit and look at a product developed for the healthcare industry to help solve one of the largest issues that face doctors and hospitals: lost revenue.

CHAPTER 11

Wireless Case Study: Healthcare Charge Capture

This chapter covers a particular aspect of healthcare that impacts almost every single healthcare worker: charge capture and reimbursement. For Medicare, Medicaid, and managed healthcare organizations to be reimbursed for services, healthcare workers must use an alphanumeric system to translate specific medical diagnoses and the related procedures to fix them into codes that insurance companies accept. Each diagnosis code is paired to a procedure code. Because of the sheer number of these codes (over 8,000 procedure codes and 15,000 diagnosis codes) and the rules involved with selecting them, healthcare workers find this to be a tedious and frustrating aspect of their jobs. More frustrating than their frustration is the statistic that on average each doctor in the United States loses $50,000 to $60,000 in annual revenue due to incorrect coding.

To get a sense of the challenges doctors face with charge capture and proper billing, Dr. Charlie Koo, CEO of the iMedica Corporation, reports on www.entnet.org as reprinted from the *Medical Group Management Journal*:

"... *In 1997, HCFA coding became more complex with the introduction of new subcategories unfamiliar to most physicians. To capture* *the required information, many doctors were forced to follow a foreign and time-consuming charting process. Some began to lose hundreds of dollars a day by chronically under-coding to simplify the process and avoid potential fines for incorrect coding . . . Many physicians hired external contractors or internal administrative staff to review charts and determine HCFA coding levels. At the same time, coding experts urged doctors to become familiar with CPT and ICD-9-CM books that, pound for pound, rival encyclopedias. These manuals contain hundreds of pages of technical terminology that physicians must use to receive proper reimbursement from payers . . . Yet regulators are fining small-practice doctors and will continue aggressively to pursue practices of all sizes . . . For example, in 1999, a Cincinnati physician was fined $242,103 for improper billing, up-coding, and related activities. In 1998, an Altoona, PA, doctor was fined $300,000 for submitting claims incorrectly. A Needham, MA, physician was fined $90,000 for submitting claims not documented in the patient record . . . "*

In addition, absurd and complex rules make charge capture even more complicated than wading through the thousands of available codes. For example, some procedure and diagnosis codes that you would think could go together cannot. Let's say that a patient visits an Ear, Nose, and Throat (ENT) doctor with the symptom of loss of hearing. After an examination, the doctor determines that the two procedures of a simple ear canal cleaning and a follow-up audio test would be the recommended course of action. According to coding regulations, the audio test and cleaning cannot occur during the same visit if a doctor wants to be reimbursed for both. The cleaning must occur during one visit and the audio test at another. This is an obvious inconvenience to the patient and the doctor.

Ultimately, coding pairing, ordering, and so on is necessary to prevent fraudulent claims against insurance companies. Some of these claims are done out of malice and some out of ignorance. Either way, healthcare professionals face lost revenues and potential fines for incorrect charge capture. So all of this complication leads doctors and others to take classes costing hundreds or thousands of dollars to learn to code correctly and accurately at the highest reimbursement levels.

Many software companies have created systems to assist healthcare workers in choosing codes so that many of these problems can be lessened or avoided altogether. This case study looks at a particular software product developed for Curon Technologies (www.curontech.com) to solve reimbursement issues and enable doctors to capture this lost revenue using a handheld device.

ICD-9 and CPT Background

The crux of the product is based on the determination and selection of the two types of codes described previously. Specifically, the first code type is dubbed a Common Procedure Terminology (CPT) code. CPTs cover any type of healthcare procedure that is rendered for a patient. Although the terminology is much more technical, procedures could include fixing a broken bone, excision of a mole, or administration of a certain drug. The second code is called the International Classification of Diseases, 9th Revision (ICD-9). (These will be replaced soon by ICD-10 codes.) As its name implies, a healthcare worker can look up any possible diagnosis code for any type of problem a patient might have: diabetes, lung cancer, and so on, but at a much more technical and specific level. CPT and ICD-9 codes are then paired together. Table 12.1 lists some example CPT and ICD-9 codes.

The World Health Organization (WHO) published the original version of ICD codes in 1948 to track morbidity and morality rates across the globe. The goal was to compile statistics on general global health

Table 12.1   Examples of CPT and ICD-9 Codes

CPT	CPT DESCRIPTION
25075	Excision, tumor, forearm, and/or wrist area; subcutaneous
25076	Excision, tumor, forearm, and/or wrist area; deep, subfascial, or intramuscular
25077	Radical resection of tumor (such as malignant neoplasm), soft tissue of forearm and/or wrist area
ICD-9	**ICD-9 DESCRIPTION**
171.2	Mal neo soft tissue arm
215.2	Ben neo soft tissue arm
238.1	Unc behav neo soft tissue
239.2	Bone/skin neoplasm nos

trends by enabling people and governments to diagnose diseases consistently. Over the next few decades as this international classification procedure gained in popularity, the National Center for Health Statistics in the United States modified and used the 9th Revision of ICD codes as a basis for collecting clinical information on patients beyond basic statistics. (*Source:* Beacon Healthcare Solutions)

The Health Care Financing Administration (HCFA) requires both CPTs and ICD-9s for reimbursement. HCFA is responsible for the Medicare and Medicaid insurance plans in the United States. (*Source:* American Medical Association). Also, insurance companies use the HCFA guidelines to determine payment(s), and it is imperative that those who bill abide by these guidelines in order to get reimbursed for the work they perform.

Solutions for Healthcare

Due to the large dollar values involved with coding, a number of companies have entered this space with solutions. They include training classes, categorized guides, Web-based solutions, PC-based software, and handheld software. This chapter is concerned only with the handheld solutions for charge capture. Out of all of the solutions a handheld solution, if designed correctly, promises to be the best one. Healthcare workers would be able capture the diagnoses and procedures at the time and place they do their work. For example, surgeons can dictate procedure information while they are performing surgery. Also, doctors could record diagnosis and procedure information while making rounds within the hospital. The location sensitivity of this information is high for a healthcare worker's business process.

MD Coder was designed and developed during the summer of 2000. A plastic surgery resident thought that a lot of time could be saved and dollars recouped through the use of handheld technology for charge capture. Her research of the market found a number of handheld applications to accomplish this task. Unfortunately, her review of them found little creativity in their design to speed up the process of finding and selecting any one of thousands of codes. They had one or more of the following problems:

- The solution offered no synchronization to a desktop computer. The handheld acted as a simple reference guide. Once you found the code you wanted, you copied it to the paper form you needed. In this respect, the handheld solved the problem of carrying and referencing large guides, but little else. The solution did not leverage some of the abilities in synchronization of delivering captured directly to a billing system.

- The CPT and ICD-9 codes were shown as long (and I mean really long) lists of codes sorted in some order. Users either scrolled through the list to find the code they needed or clicked on a Windows Explorer-like tree structure to find the appropriate code. Either of these solutions did little to speed the location of codes.

- CPT and ICD-9 codes were not associated to a particular patient, making patient tracking difficult.

Solution Design

Subsequently, our client came to us with the following needs for a handheld charge capture application:

- Quickly find any CPT and ICD-9 codes.
- Associate the codes with a particular patient.
- Synchronize codes back to a desktop PC.
- Ability to export data and print reports for charts or billing purposes.
- Easy and intuitive to use.
- Utilize the Palm platform due to its market dominance.
- System should integrate with existing hospital and physician group systems instead of replacing them.

This case study covers, more than anything else, how creative design makes for a good product. Many challenges presented themselves in trying to help healthcare workers find any one of thousands of codes on a handheld device. Moreover, the technology being used should not feel like technology. In other words, it shouldn't come with a steep learning curve.

Challenge One: Categorizing Codes

Far and away the largest challenge of the solution was how to locate the appropriate codes quickly, much more quickly than using the AMA's CPT guidebook. The resulting design relied on the recategorization of CPT codes into logical groupings. As you remember, CPTs describe the procedure, service, or product used to help a patient. Though it sounds backwards, doctors will frequently select the procedure as they're performing it and select the diagnosis after the fact.

The CPT codes were categorized first by specialty: Ear/Nose/Throat ENT, Vascular, General Surgery, and so on. Each type of doctor would then choose the database most appropriate to their practice. This code subsetting immediately limits the total number of codes in any one-specialty database by up to 90 percent.

Even with this first pass of data elimination, some databases still had 1,000 or more CPT and related ICD-9 codes. So, doctors within each specialty helped recategorize the codes into logical subgroupings. The subgroupings are organized in many cases by body part and procedure. For example, the higher-level categories may be labeled upper extremity and lower extremity, while the lower-level categories list procedures like excision or reconstruction. This recategorization enables doctors to find codes more logically.

Challenge Two: Code Lookup and Indexing

Two of the most common ways competitors displayed their codes on-screen was through long lists or a tree-like control. Either method resulted in a long list that scrolled off the bottom and right-hand side of the Palm's small screen. To prevent this type of navigation, doctors would start their search for a particular CPT code at the highest level of categorization (see Figure 11.1). A tap of the screen replaces this list with a short list of subcategories of the parent category. A second tap drills down another level. At any time, the doctor could press *Prev* to return up one level in the tree. The doctor continues this process until he or she reaches a list of specific CPT codes containing the particular code he or she was looking for.

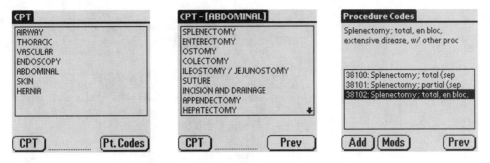

Figure 11.1 Categorized CPT codes within MD Coder.

Once the correct CPT is located, the doctor must choose a related ICD-9 or diagnosis. Figure 11.2 shows a list of ICD-9s related to the chosen CPT. Part of the code indexing process is the cross-referencing of procedures to those diagnoses that a healthcare worker is allowed to choose. This greatly simplifies choosing an ICD-9 once the CPT is found.

Okay, so now we've created a subset of categorized codes, recategorized them, and indexed them to associate CPTs and ICD-9s together. Even with all of this work, thousands of records are still under consideration. When you design for a small form-factor devices with limited resources, you must design the database to deal with these limitations. Palm-OS-based devices today are not particularly powerful devices. They have slow central processors. Because of the format, the databases are highly optimized to present codes in this drill-down fashion. If we had relied on simple searches and sorts, the waits would have been unbearable and unusable by doctors. If the doctor perceived any slowdown or wait between screens, the application would be of little use.

Challenge Three: Intuitive Interface

Doctors should feel at home within the application. As is common in many professions, doctors have little time for reading instruction manuals. Therefore, an intuitive design of interface was needed. Because patient tracking was a key requirement of the system, we sought a way to collect this information that was like something else the doctors

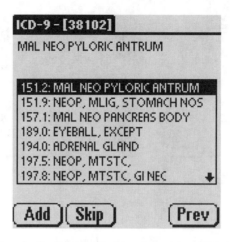

Figure 11.2 Cross-referenced ICD-9 codes based on the selected CPT code.

would be familiar with. In this case, we chose the Palm's resident address book.

The information collected for a contact in the address book is similar to the information collected by doctors about a patient (see Figure 11.3). Fields like city, state, and zip have been replaced with fields like medication, room number, and date of arrival. If you are familiar with the address book, learning MD Coder's patient tracking abilities is not a stretch.

Additionally, one other feature was added. Many of the fields like hospital and assistant are editable pick lists. Doctors can quickly create lists of common hospitals and assistants within the application and choose one from the list. The choice is added to the edit field on the right half of the screen.

Fulfilling Other Requirements

All of the information collected within MD Coder is transferable to a companion PC-application that comes with the product through a Java-based synchronization conduit. Doctors can create patients either on the handheld or the PC. Likely, the PC is the interface of choice for building a patient list due to the limitations of entering large amounts of data on handheld devices.

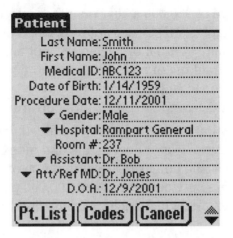

Figure 11.3 Design similarities between MD Coder and the Palm address book.

From the desktop application, doctors can create and edit patients, create and edit visits, transfer patients to and from handhelds, and do some basic reporting of patient codes. Our client knew that many hospitals have their own patient tracking and billing systems. Because of this, they knew MD Coder would not replace those systems. Rather, it would be a mobile companion to one. For those organizations that chose not to integrate MD Coder with their infrastructure, the provided PC application had enough functionality to collect and report on the captured charges.

Benefits of MD Coder

Why do doctors choose MD Coder? We think there are many reasons:

- A healthcare worker can find a unique CPT and cross-reference it to an ICD-9 in just a few seconds. This favorable result is a combination of business process and technology. Doctors in each specialty create the categorization of codes. The unique database format enables a single CPT code to be found quickly. It is much faster than using a CPT/ICD-9 book.

- Once a healthcare worker captures a code, it can be sent to the particular insurance company for payment. According to Hambrect and Co's *The Cure is in the Hand*, a physician in the United States loses an average of $50,000 a year on either miscoded charges or under coded charges. MD Coder not only helps physicians capture the patient charge with more accuracy, but enables them to have more accurate tracking of codes. If an insurance company refuses a charge, the physician is now empowered to have a detailed record of the patients and the procedure performed.

- Transcription errors are greatly reduced. Instead of relying on a hodgepodge of paper documents and index cards, the codes may be captured at the point of care at the time of care. It's estimated that somewhere between 50,000 and 100,000 patients die each year from medical errors, which include transcription errors and misread or illegible notes.

- Captured codes and patient information can be synchronized to a desktop PC for archiving, editing, and reporting.

Future Versions

During the writing of this book, a more scalable enterprise version of MD Coder was underway for a hospital group that wanted to use the software with their physician network and residents alike. The database has been moved off the PC and moved to an enterprise server. The Access database has been scaled up to Microsoft SQL Server to handle hundreds of simultaneous connections. The hospitals' patient tracking system acts as the source of patients' demographics for doctors.

Also required is wireless middleware. The original Java conduit has also been rewritten for a middleware server. Because most synchronization servers support standard IP connections, doctors have multiple choices for synchronizing patient and charge capture information: cradles, network cradles, infrared, wirelessly, and remote dial-in. Security is enhanced through authentication and encrypted connections between the handheld and wireless middleware.

The Bridge from Here to There

Healthcare is obviously an excellent market for handheld applications. Numerous companies are rushing to replace paper processes with electronic ones at the point of care. Some of the other areas where hand-held and wireless technology can help healthcare include prescriptions, physician orders, assisted living, home healthcare, hospital rounds, continuous medical education, emergency medicine, lab results, supply management, and inventory management. Like any successful wireless application in any vertical, it must prove to be intuitive, efficient, and effective in simplifying the work processes of the environment.

I hope you found the previous two chapters interesting, useful, and more importantly *tangible*. We took customer needs and designed system features to support these needs. Handheld technology was an obvious choice because of the time and location-sensitivity of the business processes. The two scenarios, hospitality/B2B and healthcare, couldn't be more divergent in background. However, in both cases, handheld and wireless computing solved problems because of time and locations natures of the business processes. In the final chapter of the book, we'll depart from today's technology and business plan and take a gander out to the near future for new (and exciting) technologies that have a high percentage chance of impacting your business.

CHAPTER 12

Wireless Technology Trends and Conclusions

I've been in the computer field consulting, writing, teaching, and programming for 23 years. I learned to program BASIC at age 7 and C at age 13. For better or worse, this is the only field I can see myself in for the simple fact of how quickly technology changes. After all this time, I continue to be blown away when attending trade shows and reading publications about technology. Just when you think that graphics can't get better, that processors can't go faster, that they'll never be able to interface a car and the Internet, and that handhelds can't be more innovative or powerful, some company surprises you.

As each of these technologies cross your desk, differentiating between fad and business trend with value always presents a challenge. There are 20 or 30 technology trends that *could* impact enterprise wireless computing. This chapter focuses on near-term trends and technologies that are *likely* to impact wireless and handheld computing. These impacts may reduce its costs, increase its performance, or generate new opportunities for your business.

Device Convergence

The first trend to impact your business will likely be device convergence. Device convergence was talked about long before handhelds and cell phones came on to the scene. In the mid-90s, hundreds of news articles swirled around the concept that television and computers would combine to become one information delivery tool. If you think about it, a television program is simply another type of transmittable data. Companies like Gateway created the first version of these convergence devices, incorporating a computer and a television with a wireless mouse. Web TV is also an attempt at device convergence. These ideas have been met with limited success. One of the challenges is marrying a highly interactive past time (computing) with an almost completely passive one (television).

In the wireless space, the belief is that handheld devices and cell phones will converge into a single device. The Samsung SPH-I300 in Figure 12.1 is one of many devices being released to deliver on this promise. This particular smart phone incorporates the Palm OS platform, but you will find other handset manufacturers incorporating the Windows CE and Symbian platforms as well. These new devices combine the best of both worlds and integrate the functions of both the traditional handheld and cellular phone. Done correctly, convergence devices will likely avoid the demise of the computer/television mentioned earlier.

The handheld device gains a wireless connection, and the cellular phone acquires the ability to manage personal information as well as enterprise information. Most mobile professionals, if not all, carry both types of devices. Combining the two into one only makes sense. You can use the address book of these devices to quickly place calls by tapping the screen. Caller ID, based on the address book, provides the information on the person calling. Some devices will enable for you to scribble a fax, tap the fax number, and send it. Outside of these integrated features, you also have all of the wireless data applications you could want: Web browsers, WAP browsers, e-mail, shopping, news, sports, and so on. Provided the manufacturers can deliver on size and battery life expectation, they will likely have a winner.

Figure 12.1 The Samsung SPH-I300 smart phone.

By the time this book lands in stores, there should be a number of smart phones similar to the I300 available on the market, including Kyocera's second foray into this space, the QCP-6035. On the flip side, manufacturers like Handspring are adding cellular phones to their handheld devices. Handspring offers a GSM phone module that plugs in as a Springboard. Either hold the PDA up to your face, or, more likely, use an earpiece and microphone.

Both Samsung and Handspring are trying to do the same thing, but from opposite angles. For some period of time, many of the lower-end phones will only inherit a WAP browser for data communication. High-end phones in the short-term will likely gain a platform like Palm, Windows CE, and Symbian. Because handhelds are slowly evolving to wireless out-of-the-box devices, it makes sense to throw in voice communication as well.

Business impact: Convergence will result in fewer devices per worker, less capital costs, and more application flexibility.

XML and a Future Wireless Architecture

The eXtensible Markup Language (XML) is a key technology not only to the Internet and business in general, but to handheld and wireless devices of all stripes. We touched on XML briefly at different points within this book, but never spent any time to really understand what its role is in wireless. It is worth spending this time with XML to understand what it is and how it could and will likely affect the overall wireless architecture we set forth in Chapter 4.

XML, still in its original 1.0 version, is a markup language in much the same way HTML is a markup language. If you've never seen HTML, you can think of a markup language as a text file that has a certain format and layout that can be interpreted by a computer application. In the case of HTML and XML, *tags* are one of the fundamental items you'll find in these files. Tags are placed around information to help describe that information. The following is an example of a tag:

```
<ADDRESS>
```

In XML, all tags have a beginning tag and an ending tag to define the limits of the data described within. An end tag looks like this:

```
</ADDRESS>
```

Between these tags, we place the actual information we are tagging as the address:

```
<ADDRESS>123 Main Street</ADDRESS>
```

To extend our example slightly, one can create a file that describes a person:

```
<FIRSTNAME>Scott</FIRSTNAME>
<LASTNAME>Sbihli</LASTNAME>
<ADDRESS>123 Main St.</ADDRESS>
<CITY>Cincinnati</CITY>
<STATE>OH</STATE>
```

XML and its capabilities are much more extensive than this simple explanation, but it suffices for our discussion. Now we have a very generic way to describe data and much more importantly we have a common format for *sharing* it between different systems. That is, we can share it provided these tags have some commonality between them. If your company chooses to describe a customer's first name with the tag

```
<FIRSTNAME>Scott</FIRSTNAME>
```

and a partner company you share the information with is expecting

```
<FNAME>Scott</FNAME>
```

You will have issues with completing the transaction because the receiving computer will likely generate an error.

One other point worth making is that XML does not have any predefined tags. The set of tags you use in an XML document is made by whomever needs to make that XML document. Therefore, it follows that XML without standard tags is worthless; different systems would not be able to share data. Thankfully, numerous organizations from various industries are working to create XML standards that all companies within a particular industry share. For example, the organization Health Level 7 (HL7) is a not-for-profit organization with the following Mission Statement:

> "To provide standards for the exchange, management and integration of data that support clinical patient care and the management, delivery and evaluation of healthcare services. Specifically, to create flexible, cost effective approaches, standards, guidelines, methodologies, and related services for interoperability between healthcare information systems." (Source: www.hl7.org.)

One of the things they're accomplishing is coordinating with healthcare providers all over the world to create XML-based data interchange standards for all types of healthcare data like patient information, doctor's orders, and lab results. If the challenge of a common format for medical data can be overcome, the sharing of information for billing, reimbursement, and change of provider activities would occur much more smoothly. If for example, you went to Switzerland to go skiing and broke your leg, in theory a doctor there could request your medical history from your personal doctor without regard to what type of computer system she has. HL7's XML-based formats would permit this. I will not get into patient confidentiality and security issues. That is a different topic covering technology, politics, and ethics. However, one can easily see the benefit of sharing data that was difficult to share previously.

With this flexibility in sharing data comes a cost: wordiness. As you can see previously, XML is a text-based format, which means that XML files take up a large amount of space compared with binary files. Large files imply longer transfer times wirelessly. Wireless networks will need to take this into account by somehow guaranteeing that whatever XML is sent and received is in a compressed format to save time and space.

Content and Presentation

In our previous example, XML described the data (the content) itself, but nothing in that description states how that information should be displayed on a computer screen in a Web browser (the presentation). For example, should the name be shown **bold** or *italics*? Should the address information be left justified or centered? How this data is visually rendered on the screen is, again, its presentation. HTML, in contrast, contains both content and presentation information in one file. HTML would contain the address of the person described previously as well as what font to display it in, what size to show, and so on.

I'll cut to the chase; this is a very bad idea. If you embed both the content and presentation of that content in one file, that file can only be used with one type of computer with one size of monitor. If you've ever seen a regular Web page like www.cnn.com within a Web browser on a handheld device, you'd understand what I'm talking about. Everything is smashed together. Tables don't line up, frames aren't handled properly, graphics cover the entire screen, and so on. Part of the XML standard is the separation of presentation and content.

Part of the XML standard are eXtensible Style Language Transformations (XSLT) documents to transform XML data into something that can be presented on a particular display. In a nutshell, you combine the content of the XML data with the presentation information in the XSLT document to show the information on any device whether it's a desktop computer, handheld, or cell phone (see Figure 12.2). To understand XSLT and how it relates to the presentation of XML data, let's look at data flow through this generic system:

1. Some device connects to the XML Application Engine (XAE).

2. The XAE detects and remembers what type of device has connected and also what the request for data is.

3. The XAE queries a corporate database for information and has returned to it a generic XML data set. This data has no presentation information. It is simple, raw data in XML format.

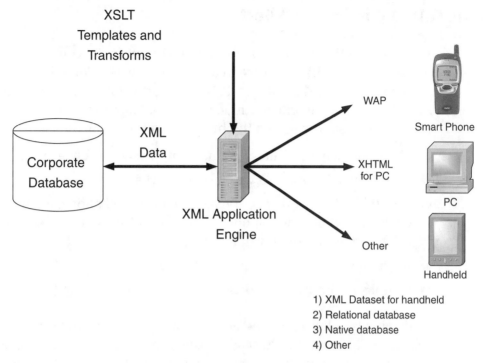

1) XML Dataset for handheld
2) Relational database
3) Native database
4) Other

Figure 12.2 XSLT adds presentation and transformations to XML data.

4. The XAE applies an XSLT template to the data based on the type of device and format it's needed in. If a WAP browser from a smart phone connects, for example, the XAE would apply the smart phone WAP XSLT template. That template would likely strip out the majority of the XML data due to the smart phones limited display size and transform the data into a WML document. Remember, WML is WAP's markup language. Similarly, if a PC running Netscape connected, the XAE would apply an XSLT transform suitable for a PC display. This would likely include retaining all of the XML data, translating it to HTML, and adding graphics. Finally, the XML data could be converted into an Access database for delivery to a PocketPC.

5. The transformed XML data is sent back to the device for display in a Web browser, WAP browser, native application, or some other type of application.

The Future Wireless Architecture

Because of XML and other factors, the wireless architecture presented in Chapter 4 will likely evolve in a number of ways. Some I believe to be more probable than others. It is also worth stating that this future state isn't an end; it will continue to evolve. The future wireless architecture I envision will need to obtain the following qualities over time:

- More reliance on wireless connectivity versus wireline
- Higher speeds for these wireless networks
- The ability for any device (desktop, laptop, cell phone, handheld, cable set top box) to connect to a back end system and receive information tailored for that device
- Minimal rework to support new devices
- Options for information delivery that include voice and video

XML will also play a key role in the evolution of the wireless architecture (see Figure 12.3). Although this picture does not portray all possibilities for connectivity, it does show computers, handhelds, and cell phones connecting through a variety of methods. Our wireless middleware has changed to some type an XAE. Today's middleware, for the most part, relies on the proprietary binary formats and APIs of the handheld to send data back and forth. Because of this, vendors must choose to support specific device types, making development time longer.

Moving from a proprietary approach to one using XML could take some time. This architecture relies on faster wireless networks, XML support with all devices, optimized messaging, and so on. However, I believe this approach or something similar will happen. There will be just too many types of devices in the future accessing the same data to make separate custom applications.

Such a system would bring immeasurable flexibility to wireless solution providers. Any time a new device or markup language is introduced, an XSLT sheet could be designed and implemented quickly to convert data into that target format. Within a much shorter period of time, that device could use the same sets of application and back end systems as the other devices.

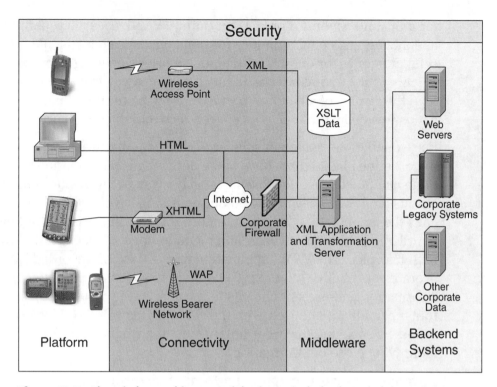

Figure 12.3 The wireless architecture of the future includes huge helpings of XML.

I admit it. That was a *huge* crash course in XML, but I believe this sheds some light on not only what XML is, but also what its capabilities are and what its future is with wireless and handheld data. Many aspects of XML have direct business benefit to you that I did not cover because of space and relevance issues. I encourage you to learn more on the topic, if not for handheld computing, then for the benefits in other computing areas. Many great books on the topic of XML exist. See the resource section at the end of this book for some of my favorites.

Voice XML

VoiceXML (VXML) is a specific implementation of the XML language definition; it is made for speech applications. In much the same way that HTML is used to markup graphical Web pages, VXML is a markup language for speech applications. HTML relies on a keyboard and mouse for input and a monitor or LCD for output. VXML uses microphones, telephones, and cellular phones for input, and speakers, telephones, and cellular phones for output. With mobile workers needing information quickly while driving in a car, taking attention away from the road to read the display of a handheld could be a recipe for disaster. Using a hands-free cellular phone with speech recognition could work well.

For example, the customer of a large parts supplier wants to know if the order he placed three days ago will reach his plant in time to continue manufacturing. Because he is on his way to the airport, pulling out his PocketPC may not be an option. Using his cellular phone, he calls the manufacturer's automated order status line. Instead of keying in order numbers, passwords, and menu options, he does all this work with his voice. The VXML application uses either pre-recorded speech prompts or robotic-sounding (today) text to speech capabilities to read information and present choices to the user of the system. This customer can identify himself with the sound of his voice and look up orders by speaking the order number, company name, or some other identifying piece of information.

Anytime you have a computer perform voice recognition, voice synthesis, optical recognition, and so forth, you will have some issues. Speech recognition, the key to this technology, is getting better by the year. You could classify it today as *pretty good*. As these systems learn to handle speech impairments, accents, and other variations, more companies may adopt VXML for delivering voice applications. In fact, if you think back to our discussion of XML and the future wireless architecture, VXML is just another output method. Instead of creating an HTML page, the transformation server can create a page of information that is read to the listener on the phone.

The VoiceXML Forum (www.voicexml.org) sees this technology being used in a variety of applications:

Generic information retrieval. This could be a weather forecast after stating the cities' names or local movie times. You could also use voice to retrieve delivery times of packages, listen to news or sports, or order a pizza. The uses are almost endless. Due to the variability in voices, many initial applications will prompt users for information that have a finite number of responses—for example, weather forecast retrieval by zip code. Each digit of a zip code can only have

10 responses (0 to 9) provided users don't respond with answers like, 4,200.

Commerce. Commerce could be another application with good adoption. If a customer knows what he wants to order from a catalog and he can ID himself to the company's VXML system, he could place orders by stating product numbers and clothing sizes, for example. Once the customer is recognized, the system could assume (or not) the shipping and billing addresses and verify them with the customer to expedite the ordering process.

Unified messaging. Retrieval and submission of calendar, address, to do, and e-mail information. Instead of using a handheld device, especially if it's misplaced, you could request a phone number from your cellular phone or retrieve your calendar for the day.

Enterprise applications. You could say that enterprise applications would fall under one of the previous categories, but numerous unique applications exist for businesses and their customers. Doctors could retrieve lab results, sales teams could look up customer accounts and place orders, and financial institutions could offer banking and brokerage services. Some applications will be similar to the today's familiar press 1 for . . . , press 2 for . . . applications of today. However, the data that goes in and out a VXML application would be little different than what would be delivered to a wireless handheld. If I've said it once, I've said it a thousand times: XML and its derivatives will enable one application to be supported across many platforms. Voice is a platform.

Business impact: XML promises to be a common way different systems and companies will share data. Various organizations, vertical markets, and technology are supporting its development. One way it could impact wireless is that one server-based application will be able to support any type of device platform that connects to it.

MPEG-4

MPEG stands for the Motion Picture Experts Group. They define a number of still and motion video and audio formats that many products use. For example, DVDs watched at home are encoded in a format called MPEG-2, which dictates certain qualities about the resolution, audio quality, and bandwidth needed to show a movie. The famous or infamous, depending on your perspective, MP3 format is another output of MPEG. Its full name is MPEG Audio Layer 3. In January 2000, MPEG released the second revision to their high-quality, low-bandwidth video format, MPEG-4. MPEG-4 is optimized for a variety of applications, one of which is the streaming of video over wireless networks. Compare MPEG-2, which is commonly used on desktop computers, to show video and with DVD to the MPEG-4 format in Table 12.1

MPEG-4 can store video of almost the same quality as MPEG-2, but using seven times less space. MPEG-4 has enhanced features like extended error correction to assist with the challenges of a wireless connection. MPEG-4 is also much more scalable than MPEG-2. Video streams as low as 5 kbps can be created. For comparison, most wireless networks in the Unites States operate at 8 to 19.2 kbps. One of the downsides currently for MPEG-4 is that encoding video into that format takes a huge amount of computing power. Similarly, end-users pay for this compression and quality in that the wireless device receiving the stream will also need fast processors to handle the decoding of the video. For those applications requiring streaming audio and video to a wireless, MPEG-4 looks to be a contender in the space.

Business impact: The ability to deliver business applications utilizing streaming video sooner rather than later.

Biometrics

Biometrics in the computer field is defined as, " . . . authentication techniques that rely on measurable physical characteristics that can be

Table 12.1 MPEG Comparison

ATTRIBUTE	MPEG-2	MPEG-4
Maximum resolution	1,920 × 1,152	720 × 576
Audio channels	8	8
Data rate @ 720 × 576	6.5 Mbps	.88 Mbps
Hardware encoding requirements	Medium	High

automatically checked" (Webopedia.com). Simply put, it's using personal biological characteristics like fingerprints, retinal scans, and voice recognition to identify someone as having permission to access certain data or systems. Soon the days of a million usernames and passwords will be gone. Also, IT departments will be ecstatic that their users aren't using their childrens' and pets' names for passwords. If the fingerprint or palm print doesn't match, you aren't allowed access.

For handheld devices in different fields, especially healthcare, biometrics may prove to solve some of the headaches of security. At one major medical university we worked with, the doctors loved doing patient rounds with information on a handheld device. What they didn't like was constantly having to enter a password each time they turned their device on. Security regulations require this to protect patient data. However, this step slowed access to the data, giving the impression of that of a mediocre handheld solution. A biometric device would solve this issue. A fingerprint is unique to an individual. Each time she turned the device on, a doctor could use her fingerprint to identify herself to gain access to patient information. No passwords to remember. Security levels remain intact. Different types of biometric identification and verification techniques include

Hand geometry. If you lay your hand flat on tabletop, it has a very unique shape. Not all of your fingers are perfectly straight. Fingers have bumps and bends in them as well as other unique characteristics. These features make up a hand geometry scan.

Facial recognition. Everyone's face looks different even if you swear your best friend looks like Harrison Ford. The position of the features on the face, as well as the size and shape, all lend themselves to a unique face.

Voice recognition. Voices have very unique patterns to them, even ones that sound the same.

Iris/retina patterns. "Loooook into my eyes," as Count Dracula might say. Your iris, the blue, brown, or green part of the eye, has a unique pattern. At the back of the eye, the retina also has a unique pattern. These can be used to authenticate users.

Signatures. Signatures have unique properties to them. Some of the more advanced features not only measure the layout of the signature, but the pressure and speed at which we make the individual strokes.

Fingerprints. Fingerprints are unique. We all know this. When fingerprints were first suggested as a way to enable people to have access to information or a physical space, morbid individuals exclaimed that a new wave in crime would occur. People would have their fingers cut off so the bad guys could gain access. Fortunately, the latest fingerprint identification systems can differentiate between live fingers and dead ones.

A number of manufacturers are working diligently to create handheld and wireless solutions using biometric identification. Fingerprints and retinas seem to be the most accurate way to identify people. Because of the inherent complexity of a fingerprint or retina along with invariability of these biometrics (signatures can change over time, facial features change as well), you gain a higher level of accuracy in identifying an individual. I suspect manufacturers of these types of devices will try a multitude of solutions. However, I believe fingerprints will be the most common way people will identify themselves to handheld devices as soon as issues related to battery and processing power are solved. (To quickly identify a fingerprint requires a fair amount of processing and battery power.) Due to the small size of handheld devices and smart phones, it seems unrealistic to identify a person by his hand geometry. For ergonomic reasons, I doubt a user would be willing to hold a handheld device up to his face for iris pattern recognition.

Business impact: Robust security from biometrics will protect corporate data and open up mobile commerce transactions.

Bluetooth

Bluetooth is another over-hyped technology buzz phrase that's captured the imagination of business and technologies companies alike. Bluetooth is considered a piconet technology similar to the terms LAN and WAN. Piconets are ad hoc, wireless networks set up within a small, defined area. Named after the Viking King Harald, Bluetooth was designed to unite different types of devices so that they can talk to each other and share information without the need for any type of cables.

Bluetooth is similar to WLAN technologies like 802.11b (wireless Ethernet) in that it transfers data in the unregulated 2.4-GHz frequency range. Unlike 802.11b, Bluetooth requires much less power to operate, enabling battery powered devices to have wireless communication. Additionally, the Bluetooth specification was designed so that the hardware needed to support it could be manufactured for five dollars per chip (in the future), making it feasible to most device makers. The downside is that a maximum of eight devices within a ten-meter radius can connect to each other with one device acting as the master and the other seven as slaves. Bluetooth can run in different modes, but one popular one has data going from the master to slave at around 700 kbps and from slave to master at 56 kbps. Compared to wireless Ethernet, Bluetooth seems a bit light on performance.

Okay, enough of the technical overview. Over 1,000 companies support the Bluetooth standard, helping to ensure success. Shortly, a deluge of devices will enter the market supporting Bluetooth. You will find Bluetooth supported in desktop computers so the PC can talk to a peripheral without a cable. You'll find it embedded into cars, appliances, and set top cable boxes in addition to handheld and wireless devices.

How does this affect business, specifically your business? Initially, some practical and smart ways exist. In the not too distant future, developers will invent new uses for it based on some of its unique capabilities. One of which is the ability for a group of Bluetooth-enabled devices with a small area to automatically start communicating should it be designed in the software. The example I'll pose may not be practical at first read, but I encourage you to think about it for a while.

I'm the Expert!

Imagine a Bluetooth-enabled Windows CE device has an application installed called *I'm the Expert*. Within it, you set preferences about subject areas that you are an expert in personally or from a business standpoint and areas that you are looking for expertise. So, let's say I'm an expert in wireless, handhelds, bicycling, and golf. I also set up some preferences stating that I am looking for specific expertise in biometrics (for a book I'm writing), a headhunter to fulfill a vice president opening, and a local golf team who needs a player. A smart phone on someone else's person also runs this same application. Walking through an airport, these two devices come within range of each other. If the application is set to automatically look for others, it would find out that the smart phone-toting person is a headhunter with 10 years of experience in the high tech field and is an avid golfer. I may choose to set the device to notify me that the person is near me, or maybe it just trades business card information for a later phone call. I know this must sound far-fetched considering the security and privacy issues, but some version of this application will happen. It has interesting enterprise and personal implications.

Other more down-to-earth ideas for Bluetooth that will likely result: cell phone ear pieces that communicate wirelessly with the cell phone, receiving floor plans and information when walking into malls, exchanging business cards automatically, printing documents by walking within range of a Bluetooth-enabled printer, and wirelessly sharing documents between handheld devices. In any event, the low cost and power consumption of Bluetooth technology will ensure its pervasiveness and forever eliminate many of the cables we use to connect devices together.

Business impact: More worker productivity. New applications and revenue channels.

Third Generation Wireless

Third Generation wireless or 3G, as the press refers to it, is supposedly the high-speed wireless WAN connectivity everyone is waiting for. The marketers make us believe 3G is the silver bullet that will cause the adoption rate of wireless Internet access to explode and boost mobile commerce into the hundreds of billions of dollars. Maybe it will and maybe it won't. Let's look at a few key points of 3G and what its implications are to your business.

The first key is performance. 3G is typically defined as 384 kbps performance while mobile (in motion) or 2 Mbps while in a fixed location (standing or walking). To compare 3G's performance to other things that you may be more familiar with, see Table 12.1. The last column of the table shows performance as a multiple of a standard 56k modem. As you can see, 3G promises to offer some impressive performance, which will open up a host of options for wireless service providers including audio, video, music, and movies.

The largest question people in wireless and telecommunication space are asking themselves is: Do consumers or enterprise customers care about 3G? For handheld devices and smart phones, I say no. Between

Table 12.1 Performance of Different Communication Technologies

TECHNOLOGY	SPEED	MULTIPLE OF A 56-KBPS MODEM
CDPD wireless modem	19.2 kbps	.3
PC modem	56 kbps	1
ISDN	128 kbps	2.3
DSL	1.5 Mbps (practical)	18.5
Cable Modem	2–3 Mbps (varies)	37.1–55.6
3G	128 kbps–2 Mbps	6.9–37.1

today's WWAN Internet bandwidth and 3G's lies 2.5G, which gives most of the performance of 3G on a shorter time frame. Again, wireless networks are the subject of numerous books and white papers. If you are so inclined, by all means read them. I've outlined a few in the resources section at the end of the book.

When determining bandwidth needs of different devices, you must look at the applications they run. In an enterprise setting, most computers are connected to 10- and 100-Mb LANs. Compare this to the numbers in Table 12.1, and you see that the slowest wired LAN is much faster than 3G in comparison. Desktop corporate computers need as much bandwidth as you can throw at them. Multimedia Web pages with Flash and Java require significant bandwidth, as do transferring large files to shared drives and using corporate applications.

Contrast this with a handheld device. Applications for handhelds include simple Web and WAP pages, data collection, and mobile commerce transactions. To handle these types of applications, there is no real need for 2 Mbps. Even streaming audio and video can be had at just 100 or 200 kbps. Today's 2G networks (8 to 19.2 kbps) are too slow for anyone, including handhelds. 2.5G networks (64 to 384 kbps) are the sweet spot for small form factor wireless devices. The good news is that we'll see these network upgrades much sooner than 3G. Japan already has nationwide 2.5G, Europe is getting it this year, and the Unites States should follow in 2002. 3G networks will benefit laptop computer users who need to connect to the Internet from anywhere. In that case, having a connection with corporate LAN-like speeds is desirable.

Second, look at battery life. Consumers want color screens, high bandwidth wireless, audio, and video . . . or at least that is what marketers would like us to believe. I think consumers would choose devices like these if they were available, affordable, and practical. The largest limiting factor is battery life, something many people overlook. Connecting to a 3G network and showing information on color display requires a significant amount of battery power. Talk times for smart phones based on today's battery and chip technology could reduce by an order of magnitude. If that is true, a phone with a talk time of two hours would be reduced to 12 minutes when running on a 3G network.

Third is rollout timing. 3G networks must be the most talked about vaporware in the history of computing. 3G has been a buzzword for at least five years. At the time of this writing, June 2001, NTT DoCoMo had just begun a trial rollout of its 3G network. NTT DoCoMo is the most popular cell/smart phone service provider in Japan. It's iMode Web-like portal that operates on cell phone displays is hit with the Japanese. Due to the fact that Japan has a small landmass and that cellular phone subscribers have exceeded the number of landline subscribers, Japan has a vested interest in rolling out the latest technology. The trial rollout of 3G in Japan on May 30, 2001 to 3,300 subscribers is the first of its kind. However, even with the ability to begin trials to gauge demand for the service, NTT DoCoMo delayed its nationwide launch by four months due to "slow delivery of both 3G-capable phones, infrastructure elements, and problems with the underlying software for controlling the network" (Source: allnetdevices.com, April 24, 2001 news article).

Finally, we have compatibility. 3G is a generic term, not a specific technology. Therefore, 3G is not some compatible utopian technology where any cellular phone in the world can be used anywhere else in the world. Multiple 3G wireless protocols exist. As you recall from the wireless primer chapter, the three basic cellular voice technologies are GSM, TDMA, and CDMA. Each have its own path to get from 2G to 3G and the end result is *not* a single standard (see Figure 12.4). Without a doubt, this is a complex alphabet soup, and each of three main technologies have more than one path they can take. For many years to come, smart phones will be incompatible between networks and countries with the exception of those smart phone manufacturers that make dual and tri-mode phones.

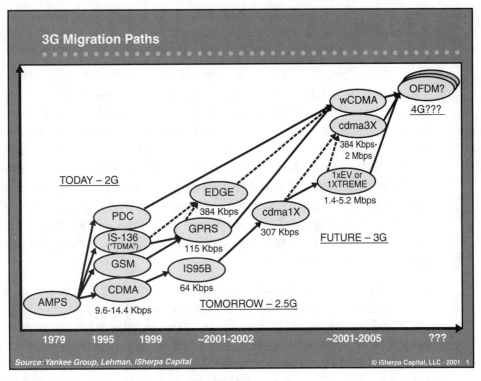

Figure 12.4 Many paths from 2G to 3G exist.

As you can probably guess, 3G is neither cheap, compatible, nor available. Do not plan your enterprise's strategy around its availability. Conversely, do not delay the rollout of applications until it is available. Today's networks work well for handheld business applications. With the imminent rollout of 2.5G technologies, businesses will have access to the sweet spot in wireless bandwidth. Within 12 months there will be a number of technologies that utilize 2.5G's bandwidth to deliver streaming audio and video as well as more complex mCommerce applications to handheld devices and smart phones.

Business impact: Not much for handheld and wireless devices. 2.5G technology is the more interesting play. 3G will just be whip cream on the sundae when and if it is available.

Voice Recognition

Voice Recognition (VR) has been around for a number of years on desktop computers. With a microphone headset, you could dictate documents into Microsoft Word or control your computer's operation. Two basic types of VR exist: discrete and continuous. Discrete means that the computer system can handle single word commands or phrases from a finite vocabulary. Delta Airlines' 800-number uses this type of technology when it commands you to press or say one. If a particular menu has five options, you can respond with five discrete, one-word answers. Contrast this to continuous speech, which is a much larger challenge. In theory, the computer system will recognize your speech as quickly as you can speak it.

These technologies, although making huge headway over the past few years, still have significant challenges ahead of them. Most continuous speech systems must be trained to learn a particular voice before it becomes accurate. This isn't particularly reassuring for a field force using handhelds that want to dictate notes into the device and have text appear on the screen. Foreign languages, accents, and speech impediments also limit the accuracy. However, most of these issues will be sorted through in the coming years.

As recognition levels and processing power increase, handheld devices will gain speech recognition technology, discrete at first and continuous later. At some point, VR will make a pretty good alternative to keyboard entry for devices. In Chapter 8, "Handheld Application Design," we discussed the challenge of getting free form text into handheld devices and smart phones. Free form text was defined as input that couldn't be simplified through the use of checkboxes, drop-lists, or something similar. For example, if a field force worker were asked to describe the condition of a house, it would be difficult to distill down a subjective description to a series of choices. A keyboard would be the best choice for input on a desktop computer. On a handheld device, this is much more of a challenge and may require the application designer to really think about how to collect the data quickly. VR may present a solution to this quandary in the future. The worker in this scenario would simply starting speaking, and the device would input the data into the correct field.

Today, very few if any handheld devices incorporate voice recognition technology. Some cell phones enable you to dial by saying a name or number. This is a first step, but a far cry from general data entry. However, VR and text-to-speech systems can be added to applications like Microsoft Exchange or Lotus Notes to enable remote voice access to e-mail, address entries, and calendar information.

Business impact: Faster and alternative access to corporate data. A wider set of enterprise applications can be designed.

The Bridge from Here to There

We covered seven technologies with high relevance to wireless and handheld computing. None of them are here yet, but should be considered best bets for what to expect in the next 12 to 18 months. Many other technologies are out there today and more in the future that will have sound and fury associated with them. In the end, they'll signify nothing. What's outlined in this chapter will likely have significant and relevant impact to the wireless field and your business. Keep your eye on them.

Conclusions

We've covered an immeasurable amount of ground in this book. Some of the chapters in here could become a book in and of themselves. In working with wireless and handheld computing, you must consider the strategy, the business process, your products, service and customers, and a plethora of technologies. Hopefully, this book has presented not only these topics at a high level, but drilled down into those areas where you wanted a more information and detail.

I do have to say that if I could give you just one piece of advice over all others related to wireless, it would be to get out there and just try it. Buy some wireless devices and give them to employees and customers. Create a quick prototype of applications you think might be a quick win and could save your company money, speed up a process, or generate additional revenue. Don't have the skills to try? Find a consulting company who can build what you want. At minimum, this book will let you filter out the irrelevant firms to find the shining stars. I've seen some great companies with amazing ideas for wireless applications, but they get too mired in politics and second-guessing to even try a prototype. Before they know it, one of their competitors has developed and rolled out their idea. Don't let that happen to you. Determine your strategy, create your ideas, prioritize them by value proposition, and get to it!

Glossary

1G The original analog cellular networks cell phones used.

1XRTT A CDMA-based 2.5G technology. It's synonymous with EDGE.

2.5G A technology step between 2G and 3G that ups digital cellular data speeds from 8 to 19 kbps to 50 to 384 kbps, depending on the specific implementation. There are 2.5G technologies for CDMA, TDMA, and GSM.

2G The first generation of digital cellular technology. It's the most commonly-used cellular technology in the United States today. There are 2.5G technologies for CDMA, TDMA, and GSM.

3G Third generation cellular promises digital cellular data speeds up to 2 Mbps.

801.11b Wireless Ethernet of Wi-Fi. It is a high-speed (2 to 11 Mbps) wireless networking technologies for indoors.

Airtime The time spent transferring voice and data across a wireless network. It refers to the time needed to send the packets and not the cumulative time.

Algorithm A set of steps for solving a problem.

Alphanumeric Any combination of characters that include numbers (zero to nine) and letters (A to Z).

Analog A signal that is continuous versus digital, which is discrete. For example, a light switch that only turns on and off is digital. One with a dial that can be continuously varied is analog.

Antenna A structure for sending or receiving electromagnetic waves.

API Application Programming Interface. A set of functions used to program a piece of software.

ASP Application Service Provider. Many definitions of ASP exist. For this book, an ASP is a company that has applications and middleware services for rent to companies who do not want to host these applications or services.

Authentication The process of verifying your identity to someone or something. Typically revealing something you know or something you are does this.

Back end system An application or database that is housed inside of a corporate firewall. Typically, a handheld connects through wireless middleware to get to the data it needs.

Bandwidth The amount of data a network, processor, and so on, can send at a time. Bandwidth is measured frequently in units like megabits per second (mbps) or kilobits per second (kbps).

Biometrics Authentication techniques that rely on measure physical characteristics. Examples include handprints, voice, retinal scans, and fingerprints.

Bluetooth A wireless networking technology made for low-power, low-cost devices within 30 feet of each other. Its original intention was to get rid of cables connecting computing devices together.

Bps Bits per second. It's a measure of speed or bandwidth. One kilobyte = 8,192 bits.

Browser An application that interprets markup languages like HTML or WAP for display to a user. Browsers are commonly used on desktop computers (Web browsers) and newer digital cellular phones (micro-browsers).

CDMA Code Division Multiple Access. A fundamental digital cellular protocol that transfers voice and data by assigning each packet a code that only the receiver of that voice or data packet can interpret.

CDMA1x A CDMA-based 2.5G technology. It's synonymous with 1XRTT.

CDMA 2000 A 3G CDMA standard for networks that started as CDMA networks. See WCDMA for a comparison.

Cell A geographic area that a cell site covers with voice and data capability.

Cell site The location where a cellular antenna is placed. It is by definition the center of a cell.

Channel A communication pathway.

Circuit switching A way of making a connection over a wireless or wireline network that establishes a dedicated connection no one else shares. Circuit-switched connections take time to establish and end.

Coverage Within the communication range of a wireless antenna to transfer voice or data packets. Lack of coverage prevents wireless communication in most of the United States.

Cradle A plastic holder for a handheld device or smart phone typically used to transfer data to a computer or network.

Digital Representing information whether it's numbers, letters, sounds, or pictures as a stream of zeros and ones.

Dual mode A cellular phone or device that has both a digital mode and a fallback analog mode when digital coverage is not available.

EDGE Enhanced Data Rates for global Evolution, a 2.5G TDMA technology.

Electromagnetic spectrum The range of frequencies that electromagnetic radiation uses.

Embedded device Incorporating a technology into a product. For example, an embedded Java device may have a burned Java VM incorporated into a computer chip.

Encryption The process of using mathematical transformations on data to prevent someone from eavesdropping or reading it. Only someone with the proper decryption key can unlock the code to get the original data back.

EPOC An operating system created by Psion. Psion creates a series of handheld devices popular in Europe.

Ethernet A local area networking technology. Most corporate networks today use Ethernet to connect computers.

Extranet An intranet that is usable by authorized outsiders.

Firewall A hardware and/or software combination that protects a company's internal network from the public Internet by preventing outsiders from having access to internal corporate information.

Frequency The number of times some event occurs in a given interval. Used typically to describe the number of times per second an electromagnetic wave oscillates.

Gateway A hardware and software combination that connects two different networks.

GHz Gigahertz. One billion cycles per second.

GPRS Global Packet Radio System. A 2.5G networking technology.

GPS Global Positioning System. It is a system of satellites that broadcast their location. When used with GPS receiver, someone can determine his or her location on Earth to within a few meters.

GSM Global System for Mobile Communication. GSM is a 2G digital cellular technology most popular in Europe.

Handheld A computing device that fits in a shirt pocket with an operating system and the ability to connect to other devices whether wirelessly or through a wireline.

HDML Handheld Device Markup Language. The predecessor of WAP's Wireless Markup Language (WML) created by Openwave.com.

Hertz (Hz) Cycles per second.

HTML HyperText Markup Language. A scripting language used to describe data and the presentation of that data. HTML is typically shown in a browser.

Http HyperText Transfer Protocol. The protocol HTML files are transferred over.

IEEE Institute of Electrical and Electronics Engineers. IEEE is an organization that developes many computing standards.

IMode An extremely popular content portal based on wireless Web technology in Japan. IMode is owned by NTT Docomo, who plans on bringing iMode to other countries including the United States.

Infrared A set of frequencies in the electromagnetic spectrum. Infrared is used by many handheld devices to transfer data wirelessly over a short distance.

Infrastructure The underlying software and hardware required to create a software solution.

Internet A global, public network consisting of millions of computing devices.

Intranet A network that performs similar functions to the Internet, but is private to a company or organization.

IRDA Infrared Data Association. IRDA is an organization that sets standards for the use of infrared frequencies to transmit data wirelessly.

Java A programming language created by Sun Microsystems commonly used to create programs that work on multiple platforms.

JavaScript A scripting language created by Netscape to add some programming functionality within Web pages.

Java VM Java Virtual Machine. A piece of software that translates Java instructions into instructions a specific computing platform can understand.

Kbps Kilobytes per second. One Kilobyte = 1,024 bytes.

kbps kilobits per second. One kilobits = one-eighth of a Kilobyte.

KHz Kilohertz. One KHz = 1,000 hertz.

Latency The amount of time spent waiting for something to happen.

LAN Local Area Network.

Mbps Megabits per second. One megabit = one-eighth of a megabyte.

Messaging middleware A type of middleware whose main function is to pass messages, which contain data, between two applications. This type of middleware has certain advantages in a wireless environment.

MHz Megahertz. One MHz = 1,000,000 hertz.

Microbrowser A browser that runs with minimal memory and processing resources available. Microbrowsers typically refer to browsers built-in to cellular phones.

Middleware Translation software that enables applications to communicate regardless of platform, standards, or vendors.

Mobile IP A standard for enabling wireless devices to deal with challenges of having a data session or phone conversation while on the move.

Modem Modulator-demodulator. It's used to translate digital information into analog signals for transmission to another computing device.

Modulate To change or vary some property.

MPEG-4 Moving Picture Experts Group. MPEG-4 is standard for streaming high-quality video over wireless (low-bandwidth) connection.

Native application Applications built to run on a particular platform or operating system. They are not Web based.

Operating system Software that manages a computing device and its resources. Sometimes used synonymously with the term *platform*.

OSI reference model A standard framework for creating communication protocols in seven software layers.

Packet switching A technology for enabling data from multiple sources to multiple destinations to share a single communication channel. *See* Circuit Switching.

Palm One of the four major handheld platforms.

PDA Personal Digital Assistant. Used as a synonym for handheld.

Peer-to-Peer In the case of wireless and handheld computing, it enables two handheld devices to communicate directly without the need for an intermediate computer, server, or piece of middleware.

Piconet A small, ad hoc network created by two wireless devices within in a close range of each other.

PIM Personal Information Manager. A program that manages information like contacts, appointments, e-mail, and to do lists.

Platform A combination of software and hardware that applications can be designed and built again. Typically, it's used synonymously with operating system.

Protocol Generically, it's an agreed-upon format for doing something. In this book, it's an agreed-upon format for transferring data over a network.

PocketPC Handheld devices that run version 3.0 of Windows CE.

RAD Rapid Application Development. RAD packages are tools for developing applications that sacrifice performance for ease of development.

Real time Systems designed to submit and retrieve data at the moment it's input or requested. *See* Store-and-Forward.

RIM Research In Motion and the maker of one of the four major handheld platforms.

Runtime A piece of software used typically with RAD tools that translates the instructions a RAD application creates into ones that the operating system can understand.

Smart phone A cellular phone with a built-in device platform like Palm, Windows CE, RIM, wireless Web, or other operating systems.

Store-and-forward A synchronization methodology by which you store changes on the device for updating at a later time. This is typically used when wireless coverage is unavailable for a real time connection. *See* Real time.

Stylus A plastic, pen-like instrument used to interact with a handheld device's screen.

Symbian A company that makes a handheld and smart phone platform by the same name. It is based on the EPOC operating system from Psion.

Synchronization The process of ensuring data in two or more data sources has exactly the same information.

SyncML An XML-based standard for synchronizing data between disparate systems.

TCP/IP Transmission Control Protocol/Internet Protocol. The protocol that the Internet (and most other networks) use to transfer data.

TMDA Time Division Multiple Access. A fundamental digital cellular protocol that transfers voice and data by assigning each voice packet or data stream its own time slot.

TP monitor Transaction Processing Monitor. A type of middleware that ensures that transactions are processed in a reliable timely manner with support for commit and rollback functions.

VoiceXML An XML-based standard that uses voice as its input and output methods.

WAP Wireless Application Protocol. A series of technologies for delivering Web-like information to small form-factor, low-powered devices. Consider WAP to be the little brother of Web-browser technologies.

WCDMA A third generation CDMA standard for networks that started as TDMA (and GSM) networks. *See* CDMA 2000 for a comparison.

Web Shortened term for the World Wide Web (WWW).

Web application Wireless applications that rely on Web technologies like HTML, JavaScript, and WAP.

Wi-Fi The common name for wireless Ethernet or 802.11b.

Windows CE One of the four major handheld platforms created by Microsoft.

Wireless A term to describe a method of communication. Computing without cables. Sometimes used to describe handheld computing whether or not it's truly a wireless solution.

Wireless device A smart phone, PDA, pager, messaging device, portable computer, or embedded device that sends and receives information wirelessly.

Wireless Internet A term used to describe the process of making WWW content available wirelessly on handheld devices. Considered, by this book, as one of the four major platforms.

Wireless middleware A version of middleware that enables back end systems and a wireless device to exchange data or two wireless devices.

Wireline The opposite of wireless. Using some type of physical cable whether it's a network cable, phone cord, or something else to establish a connection.

WLAN Wireless Local Area Network. Used to describe indoor, wireless networks.

WML Wireless Markup Language. The markup language of WAP that is similar to HTML.

WMLScript A scripting language to add programming functionality to WML pages. This is similar to JavaScript in the WWW area.

WWAN Wireless Wide Area Network. Outdoor, wireless networks covering vast geographies.

XHTML The successor to HTML that is based on XML.

XML eXtensible Markup Language. A standard specification used to describe data and to facilitate the sharing of data between different back ends.

XSLT eXtensible Stylesheet Language Transformations. XSLT is part of the XML standard that enables you to add presentation information to XML and transform it into other types of information.

Zaurus A handheld platform popular in Asia.

Recommended Readings and References

Books

Collins, James C. 1997. *Built to Last: Successful Habits of Visionary Companies*. HarperBusiness.

Giguere, Eric. 2000. *Java 2 Micro Edition*. John Wiley & Sons.

Hamel, Gary, and Prahalad, C.K. 1996. *Competing for the Future*. Harvard Business School Press.

Hammer, Michael, and Champy, James. 1994. *Reengineering the Corporation: A Manifesto for Business Revolution*. HarperBusiness.

Kaplan, Robert S., and Norton, David P. 2000. *The Strategy-Focused Organization: How Balanced Scorecard Companies Thrive in the New Business Environment*. Harvard Business School Press.

Porter, Michael. 1998. *Competitive Strategy: Techniques for Analyzing Industries and Competitors*. Free Press.

Sharma, Chetan. 2001. *Wireless Internet Enterprise Applications*. John Wiley & Sons.

Stern, Carl W., et al. 1998. *Perspectives on Strategy: From the Boston Consulting Group*. John Wiley & Sons.

Treacy, Michael, and Wiersema, Fred. 1997. *The Discipline of Market Leaders: Choose Your Customers, Narrow Your Focus, Dominate Your Market.* Perseus Press.

Magazine Articles

Brewin, Bob. October 20, 2000. "UPS Moving to Install Wireless LANs at All Delivery Hubs." *Computerworld.*

Eisenhart, Mary. February 2001. "Finding the Right Wireless ASP." *mBusiness.*

Frimmer, Justin. February 2001. "3G's Coming Energy Crisis." *mBusiness.*

Frimmer, Justin. March 2001. "XHTML: A Large Language for Next-Gen Devices." *mBusiness.*

Gibbons-Paul, Lauren. March 15, 2001. "Wireless Early Adopters: Survival Tips from the Pioneers." *CIO.*

Greengard, Samuel. April 2001. "Testing the Usability Factor." *mBusiness.*

Hess, Ed. July 2001. "Head of the Class." *Integrated Solutions.*

Kuchinskas, Susan. March 2001. "WAP 2.0: The Next Generation." *mBusiness.*

McGarvey, Robert. April 2001. "Solving the Text-Input Problem." *mBusiness.*

Petersen, Karen E., and Scarborough, Caroline. March 15, 2001. "FedEx: On the Move with Wireless Technology." *10Meters News Report.*

Rosenblum, Joseph. July 1, 2001. "Midnight Express." *Inc.*

Semilof, Margie. June 2001. "Wireless Takes to the Trenches." *mBusiness.*

Web Articles and White Papers

AllNetDevices Staff. April 24, 2001. "DoCoMo Delays 3G Rollout." www .allnetdevices.com/wireless/news/2001/04 /24/docomo_delays.html.

AllNetDevices Staff. December 28, 2000. "2.5G to Dominate. Not 3G." www.allnetdevices.com/industry/market/2000/12/28/25 g_to.html.

Andersson, Christoffer. "What Can We Learn from Japan?" WirelessDevNet.com. www.wirelessdevnet.com/GPRSand3Gapps/ articles/learnfromjapan.html.

Brain, Marshall, and Harris, Tom. "How GPS Receivers Work." www
.howstuffworks.com/gps.htm.

Camacho, Carlos. January 24, 2001. "Location-Based Services in Japan."
www.allnetdevices.com/wireless/opinions/2001/01/24/location-based_
services.html.

Carrollo, Kelly. February 26, 2001. "Eye on the Enterprise." Telecomclick
.com.

Cassel, David. March 13, 2001. "FedEx Expands Wireless Tracking."
MbizCentral. www.mbizcentral.com/story//MBZ20010312S0015

Chambers, Steve. "XML Gives Voice to New Speech Apps." NWFusion
.com. www.nwfusion.com/news/tech/2001/0730 tech.html.

Coo, Charlie, PhD. 2001. "Automated HCFA Coding an Rx for Physician
Headaches." www.entnet.org/Bulletin/hcfa.html.

English, David. January 9, 2001. "No Strings Attached." www.zdnet.com/
products/stories/reviews/0,4161,2671660,00.html.

Eurotechnology. Ongoing. "The Unofficial Independent iMode FAQ."
www.eurotechnology.com/imode/faq.html.

Gorney, Mark. 2000. "Make No Mistake about It." www.thedoctors.com/
Resources/TDA/archives/2000/tda200.htm.

Gupta, Puneet. "Mobile Wireless Communications Tomorrow."
AllNetDevices.com. www.wirelessdevnet.com/channels/wireless/
training/mobilewirelesstomorrow.html.

Haskin, David. "Has Palm Cracked the Golden Egg?" AllNetDevices
.com. www.allnetdevices.com/industry/reality/2001/06/25/has_palm
.html.

Haskin, David. "Java, Bluetooth On Verge of Long-Awaited Success."
AllNetDevices.com. www.allnetdevices.com/industry/reality/2001/06/
18/java_bluetooth.html.

Haskin, David. "Wireless Net: The Pain Before the Gain."
AllNetDevices.com. www.allnetdevices.com/industry/reality/2001/07/
23/wireless_net.html.

Haskin, David. March 5, 2001. "DoCoMo Passes 20 Million i-Mode Users."
www.allnetdevices.com/wireless/news/2001/03/05/docomo_passes
.html.

Miller, Craig. "Four Principles of Wireless Application Design."
Proxicom.com. www.proxicom.com/ebusiness_wire/pxcm_features/
display.asp?articleId=129.

Morgan, Bryan. "The X Factor." WirelessDevNet.com. www
.wirelessdevnet.com/channels/wap/features/xfactor.html.

Proust, Albert. November 3, 2000. "Personal Area Network: A Bluetooth Primer." www.oreillynet.com/pub/a/wireless/2000/11/03/bluetooth .html.

Reuters News Service. June 14, 2001. "DoCoMo (9437.T) to Replace 3G Phones." www.businessweek.com/reuters_market/P/REUT-PU0.HTM.htm.

Scanlon, Bill. March 19, 2001. "Security Is Key." www.zdnet.com/ intweek/stories/news/0,4164,2696791-3,00.html.

Steck, William. 2000. "Bluetooth Wireless Technology." www .anywhereyougo.com/Content.po?name=bluetooth/Intro.

Steve's Toronto Area Cellular/PCS Site Guide. November 13, 2000. "CDMA vs TDMA." http://arcx.com/sites/CDMAvsTDMA.htm.

Sun Microsystems. 2000. "Applications for Mobile Information Devices." pp 1–17. java.sun.com/products/midp/midpwp.pdf.

Sun Microsystems. 2001. "Java 2 Platform, Micro Edition-Datasheet." pp 1–2. java.sun.com/j2me/j2me-ds-0201.pdf.

Sun Microsystems. 2001. "MIDP APIs for Wireless Applications." pp 1–14. java.sun.com/products/midp/midp-wirelessapps-wp.pdf.

Symbian. Ongoing. "Technology: Glossary of Terms." www.symbian .com/technology/glossary.html.

The Shosteck Group. June 2001. "GSM or CDMA: The Commercial and Technology Challenges for TDMA Operators." www.cdg.org/ tech/shosteck/overview.asp.

Thorsberg, Frank. April 18, 2001. "Iscibe Signs for $15 Million." www .localbusiness.com/Story/0,1118,NOCITY_732564,00.html.

VanderMeer, Jim. February 1, 2001. "Will Wireless Location-Based Services Pay Off?" www.geoplace.com/bg/2001/0201 /0201pay.asp.

Völkel, Frank. September 13, 2000. "MPEG-4 - Copying a DVD Video to CD-ROM." http://www4.tomshardware.com/video/00q3/000913/index .html.

Wieland, Ken. September 2000. "What Are Location-Based Services?" www.telecoms-mag.com/issues/200009/tci/where_are_the.html.

Wilner, Michael. "High-Speed Text Entry for Handhelds." AllNetDevices.com. www.allnetdevices.com/developer/white/2001/02/ 13/high-speed_text.html.

WirelessDevNet.com. March 12, 2001. "FedEx Expands Wireless Services for Customers Through Agreement with w-Technologies."

Web Sites

In order to help you find the information you are looking for, these Web sites have been separated by chapter and then subject. If you're interested in companies making XML products, XML organizations, or XML news sites, for example, you'll be able to quickly find the site you are looking for without wading through a list of gigantic proportions. Not all chapters have recommended Web sites. Specific articles on a topic are listed above.

Chapter 5

Platforms and Devices

Handera	www.handera.com
Handspring devices	www.handspring.com
IBM Workpads	www.ibm.com
Information on Newton	www.panix.com/~clay/newton/
Kyocera smartphone	www.kyocera-wireless.com/kysmart/ kysmart_series.htm
Palm devices	www.palm.com
PocketPCs	www.casio.com/personalpcs/ www.compaq.com/showroom/ handhelds.html www.hp.com/jornada/ www.sharp-usa.com
Psion devices	www.psion.com
RIM devices	www.rim.net
Samsung smartphone	www.samsungusa.com
Sony Clies	www.sonystyle.com/vaio/clie/index.html
Symbian devices and smartphones	www.symbian.com
Symbol devices	www.symbol.com/products/mobile_ computers/mobile_computers.html

WAP phones and smartphones	www.motorola.com
WAP phones and smartphones	www.ericsson.com
WAP phones and smartphones	www.www.nokia.com
Windows CE	www.microsoft.com/mobile/ pocketpc/default.asp

Device Reviews and Roundups

Device reviews	www.zdnet.com/special/filters/sc/pda/ http://computers.cnet.com/hardware/ 0-1087.html?tag=dirwww.pencomputing.com

Chapter 6

Whitepapers and Tutorials

3G Info	www.ericsson.com/technology/3G.shtml
Bluetooth Information	www.ericsson.com/technology/ Bluetooth.shtml
Cell Phone Fundamentals	www.howstuffworks.com/cell-phone.htm
Cell Service Fundamentals	www.howstuffworks.com/cell-phone-service.htm
EDGE Information	www.ericsson.com/technology/ EDGE.shtml
GRPS Info	www.ericsson.com/technology/ GPRS.shtml
PDA Fundamentals	www.howstuffworks.com/pda.htm
Radio Fundamentals	www.howstuffworks.com/radio.htm
WAP Information	www.ericsson.com/technology/WAP.shtml
White Paper Repository	http://mcommerce.bitpipe.com/mcom_ index.jsp
Wi-Fi Standard	http://standards.ieee.org/getieee802/ 802.11.html
Wireless Fundamentals	www.howstuffworks.com/ wireless-network.htm

iMode

iMode Development	www.ai.mit.edu/people/hqm/imode
iMode Eye	www.japon.net/imode/
iMode FAQ	www.eurotechnology.com/imode/faq.html
iMode	www.nttdocomo.com
Java Mobiles	www.javamobiles.com/

Wireless Technology and Orgs

3G Organization	www.3gpp.org/
Bluetooth	www.bluetooth.org
CDMA Organization	www.cdg.org
CDMA Technology	www.cdmatech.com
International Standards Org	www.ieee.org
	www.iso.ch
Mobitex	www.mobitex.org
Standards Org	www.umts-forum.org
TDMA/GSM Org	www.uwcc.org/
WAP Standard	www.wapforum.org
Wireless Awareness	www.wirelessready.org

Smart and Cell Phone Vendors

Ericsson	www.ericsson.com
Kyocera	www.kyocera-wireless.com
Motorola	www.motorola.com
Nextel	www.nextel.com
Nokia	www.nokia.com
Samsung	www.samsung.com
Sony	www.sony.com

Data Networks

CDPD	www.www.cdpd.com/
DataTac	www.dopforum.com/
Re/Flex	www.mot.com/MIMS/MSPG/FLEX/
Ricochet	www.metricom.com/information/index.html

Other

Electromagnetic Spectrum Picture	www.lbl.gov/MicroWorlds/ALSTool/EMSpec/EMSpec2.html

Chapter 7

Middleware Providers

@Hand	www.@hand.com
Aether Systems	www.aethersystems.com
AvantGo	www.avantgo.com
BroadBeam	www.broadbeam.com
EveryPath	www.everypath.com
Extended Systems	www.extendedsystems.com
Geoworks	www.geoworks.com
Microsoft	www.microsoft.com/biztalk/default.asp
PumaTech	www.pumatech.com
Synchrologic	www.synchrologic.com
ThinAir Apps	www.thinairapps.com

Articles

Alerts	www.nettechrf.com/news_wireless_10-00.html
Enterprise Wireless	www.nettechrf.com/news_wireless_09-00.html
IT Investments	www.mbizcentral.com/story/news/MBZ20010531S0005

Middleware	www.wirelessdevnet.com/channels/wireless/features/middleware.html www.wirelessready.org/connectivity.asp
Mobile IP	www.hill.com/library/archives/mobileip.shtml
Mobile vs. Wireless	www.wirelessweek.com/index.asp?layout=story&doc_id=23222&verticalID=540&vertical=Software+Developer
OSI Model	www.its.bldrdoc.gov/fs-1037/dir-025/_3680.htm www.siu.edu/~bkearney/415/OSI_Data.htm
Peer-to-Peer	www.wallstreetandtech.com/story/wireless/WST20010501S0001
SMS	www.unstrung.com/server/display.php3?id=563&cat_id=2
Store-and-Forward	www.nettechrf.com/news_wireless_06–00.html
SyncML	www.syncml.org/technology.html
Transaction Processing	www.subrahmanyam.com/articles/transactions/NutsAndBoltsOfTP.html
WAP Middleware	www.wirelessdevnet.com/channels/wap/features/stuckinthemiddle.html
Whitepapers	www.luminant.com/luminant.nsf/cbb1317db020f78e862566a2007abecc/b6688db72605c9c686256a3a00780dd7?OpenDocument
Wireless Coverage	www.mbizcentral.com/story/news/MBZ20010601S0001

Synchronization Organizations

SyncML	www.syncml.org

Chapter 8

Native Application Tools

eMbedded Visual Tools (CE)	www.microsoft.com/mobile/downloads/emvt30.asp

Metrowerks Codewarrior (Palm)	www.metrowerks.com
Visual Studio (RIM)	http://msdn.microsoft.com/vstudio

Native Applications Tools-RAD

AppForge (Palm)	www.appforge.com
MobileBuilder	www.mobilebuilder.com
Pencel	www.pencel.com
Satellite Forms	www.pumatech.com
ScoutBuilder	www.aethersystems.com

Java

CrEme	www.nsicom.com
IBM	www.ibm.com/software/ad/vajava/
Java Mobiles	www.javamobiles.com/
Kada System's VM	www.kadasystems.com
Sun Java	http://java.sun.com/j2me

Wireless Web

Article	www.nettechrf.com/news_wireless_11-00.html www.thestandard.com/article/ 0,1902,15258,00.html
Ericsson & WAP	www.ericsson.com/WAP/
Nokia & WAP	www.nokia.com/wap/
WAP Forum	www.wapforum.org
WAP News	www.wap.com/
WAP Resource	www.wap.net/

Databases

DB2 Everyplace	www-4.ibm.com/software/data/db2/everyplace/
Oracle	www.oracle.com/ip/index.html?think9i_portal .html
Sybase iAnywhere	www.sybase.com/products/mobilewireless

Other Websites

Handheld Development Tools	www.mobileinfo.com/application_dev.htm
Microsoft Dev Resources	www.microsoft.com/mobile/developer/default.asp
Palm Dev Resources	http://goanna.cs.rmit.edu.au/~winikoff/palm/dev.html
Palm Dev Resources	www.palmos.com
RIM Dev Resources	http://developers.rim.net/

Chapter 11

Curon Technologies	www.curontech.com
HL7 Organization	http//www.hl7.org
Medical PocketPC	www.medicalpocketpc.com/
Palm	www.palm.com/enterprise/solutions/healthcare
Palm Healthcare Resources	http://palmtops.about.com/cs/medicalpalm
Palm Healthcare Solutions	http://www.palmpowerenterprise.com/issues/issue200103/healthcare001.html
PDA MD	www.pdamd.com

Chapter 12

Convergence Devices

Handspring Visor Phone	www.handspring.com/products/visorphone/index.jhtml
Kyocera Smartphone	www.kyocera-wireless.com/kysmart/kysmart_series.htm
Samsung SmartPhone	www.samsungusa.com/cgi-bin/nabc/main/samsungusa.jsp

XML

Microsoft & XML	http://msdn.microsoft.com/xml/default.asp
SOAP	www-106.ibm.com/developerworks/xml/library/x-soapbx2/index.html
VoiceXML	www.wirelessweek.com/index.asp?layout=story&doc_id=27832&verticalID=540&vertical=Software+Developer
VoiceXML Org	www.voicexml.org/
VXML Reference	www.zvon.org/xxl/VoiceXMLReference/Output/
XHTML Standard	www.w3.org/MarkUp/
XML	www.zdnet.com/eweek/stories/general/0,11011,2765997,00.html
XML Advantage	www.xmladvantage.com
XML Info	www.xmlinfo.com/
XML Org	www.xml.org
XML Standard	www.w3.org/XML/
XML Standards	www.zdnet.com/eweek/stories/general/0,11011,2764857,00.html
XML Times	www.xmltimes.com

Streaming Video (MPEG-4)

Activesky	www.activesky.com
Article	www.techtv.com/freshgear/pipeline/story/0,23008,3316898,00.html
Firepad	www.firepad.com
MPEG Homepage	www.cselt.it/mpeg/
Packet Video	www.packetvideo.com
Resource List	www.mpeg.org

Biometrics

Article	www.zdnet.com/pcmag/features/biometrics
AuthenTec Inc.	www.authentec.com

Biometrics Organization	www.biometric.org
Biometrics Resources	http://netsecurity.about.com/compute/ netsecurity/cs/biometrics1/index.htm
Fingerprints	www.fingerprint-society.org.uk/
I/O Software	www.iosoftware.com
Resource Site	www.biomet.org/

Voice Recognition

ART Recognition	www.artrecognition.com/
Article	http://news.cnet.com/news/ 0-1006-200-2125025.html http://news.cnet.com/news/ 0-1006-200-2751842.html?tag=bplst www.infoworld.com/articles/pi/xml/00/ 03 /07/000307 pivoice.xml
Dragon Systems	www.dragonsys.com
Everypath Voice Server	www.everypath.com
IBM	www-4.ibm.com/software/speech/
Lernout & Hauspie	www.lhsl.com/default2.htm
Nuance	http://nuance.com
Reviews	www.zdnet.com/products/filter/guide/ 0,7267,6000743,00.html
Speech for Handhelds	http://iwsun4.infoworld.com/articles/hn/ xml/01/01/22/010122 hnvoice.xml

Whitepapers and Tutorials

| LBS | www.nettechrf.com/news_wireless_ 01-01.html |
| LBS Fundamentals | www.howstuffworks.com/ location-tracking.htm |

News and Dedicated Sites

| AllNetDevices | www.allnetdevices.com |
| CE Windows.net | www.cewindows.net |

Handango	www.handango.com
The Gadgeteer	www.the-gadgeteer.com/
Mobile Info	www.mobileinfo.com
Palm Info Center	www.palminfocenter.com
PDA Geek	www.geek.com/pdageek/pdamain.htm
Psion Place	www.psionplace.com
Rim Road	www.rimroad.com
Think Mobile	www.thinkmobile.com
Wireless Dev Net	www.wirelessdevnet.com
Wireless Week	www.wirelessweek.com

Handheld Software Repositories

CE Archives	www.cearchives.com
C-Net	http://download.cnet.com/downloads/0-1726360.html?tag=dir
Handango	www.handango.com
Palm Gear	www.palmgear.com
Psion	www.psionplace.com
RIM	www.rimroad.com
ZDNet (Palm)	www.zdnet.com/downloads/pilotsoftware/
ZDNet (Windows CE)	www.zdnet.com/downloads/ce/